Rating and (Pocket Book

Matthew Cain Ormondroyd

Routledge
Taylor & Francis Group

LONDON AND NEW YORK

First published 2017
by Routledge
2 Park Square, Milton Park, Abingdon, Oxon OX14 4RN

and by Routledge
711 Third Avenue, New York, NY 10017

Routledge is an imprint of the Taylor & Francis Group, an informa business

© 2017 Matthew Cain Ormondroyd

British Library Cataloguing-in-Publication Data
A catalogue record for this book is available from the British Library

Library of Congress Cataloging-in-Publication Data
Names: Ormondroyd, Matthew Cain, author.
Title: Rating and council tax pocket book / Matthew Cain Ormondroyd.
Description: Abingdon, Oxon [UK] ; New York : Routledge, 2017. | Includes bibliographical references and index.
Identifiers: LCCN 2016045306| ISBN 9781138643802 (pbk. : alk. paper) | ISBN 9781315629216 (ebook : alk. paper)
Subjects: LCSH: Tax assessment—Law and legislation—England. | Tax assessment—Law and legislation—Wales. | Local taxation—Law and legislation—England. | Local taxation—Law and legislation—Wales. | Real property tax—Law and legislation—England. | Real property tax—Law and legislation—Wales. | Tax collection—England. | Tax collection—Wales.
Classification: LCC KD5729 .O76 2017 | DDC 343.4205/42—dc23
LC record available at https://lccn.loc.gov/2016045306

ISBN: 978-1-138-64380-2 (pbk)
ISBN: 978-1-315-62921-6 (ebk)

Typeset in Goudy Std
by Swales & Willis Ltd, Exeter, Devon, UK
Printed and bound by CPI Group (UK) Ltd, Croydon, CR0 4YY

This book is dedicated to Alice and
Arthur Ormondroyd, with all my love.

Rating and Council Tax Pocket Book

The *Rating and Council Tax Pocket Book* is a concise, practical guide to the legal and practical issues surrounding non-domestic rates and council tax. An essential tool for busy tax collection practitioners in local authorities and private practice, it will also be suitable for a range of non-specialist property professionals who may have to deal with rates and council tax matters as part of their practice.

This handy pocket guide is accessible to specialist and non-specialist alike, covering everything from key concepts through to liability, exemptions, procedure and completion notices. The book encompasses both English and Welsh law, and includes reference to all the relevant statutory provisions. With detailed discussion of key cases, this is a book that no one with an interest in rating and council tax should be without.

Matthew Cain Ormondroyd is a barrister specialising in this area of the law, having appeared in numerous cases in the Magistrates' Court, Valuation Tribunal and High Court.

Contents

Contents

Table of cases

Note: Cases are listed by paragraph/note number.

Table of statutory provisions

1

Introduction

1.1 This book aims to provide a clear, concise and practical guide to the law of rating and council tax in England and Wales. Liability to the taxes, in the broadest sense, and collection of them is the focus. As such the book does not deal with valuation or related matters. Nor does it deal with administration in the sense of internal administration of the billing authority or the setting of levels of taxation.

1.2 This introduction introduces the relevant legislation, explains the structure of the book and then provides a short summary of the law relating to rating and council tax, which is intended to assist those who may be unfamiliar with these areas of law to find their way around the detailed discussion contained in the following chapters.

GOVERNING LEGISLATION

1.3 The primary statute governing non-domestic rating is the Local Government Finance Act 1988, which is referred to hereafter as LGFA 1988. The other key piece of legislation from the point of view of collection is the Non-Domestic Rating (Collection and Enforcement) (Local Lists) Regulations 1989. These regulations are referred to as the NDR Collection Regulations 1989 and are dealt with in some detail in Chapter 5.

1.4 The corresponding provisions for council tax are to be found in the Local Government Finance Act 1992, or LGFA 1992, and the Council Tax (Administration and Enforcement) Regulations 1992, or CT Enforcement Regulations 1992.

1.5 All of these pieces of legislation have been amended multiple times over the years. All references in the text or footnotes to this or other legislation are to the up-to-date amended version.

STRUCTURE OF THE BOOK

1.6   The book is split into two parts. The first part deals with the law of non-domestic rates. The second deals with council tax.

1.7   One important reason for this structure is that the council tax provisions are, to a large extent, either dependent upon or derived from rating provisions. For example, the definition of a 'dwelling' (the taxable unit of property for council tax purposes) is tied to the definition of a 'non-domestic hereditament' (the taxable unit of property for rating purposes). The provisions on collection of council tax in the CT Enforcement Regulations 1992 shadow those on non-domestic rates in the NDR Collection Regulations 1989. It is therefore difficult to understand the council tax provisions without having some degree of knowledge of the rating provisions. Reasonably comprehensive cross-referencing is provided, however, for the reader who wants to focus on the sections dealing with council tax.

1.8   It should also be noted that although the systems are in some respects similar, there are several large differences between them when it comes to collection of the two taxes. Perhaps the most significant for present purposes is the structure of different routes of appeal and challenge. Decisions on liability to rates can be challenged in the Magistrates' Court, whereas the equivalent decisions in respect of council tax must be challenged by making an appeal to the relevant valuation tribunal. These are the Valuation Tribunal for England (or VTE) and the Valuation Tribunal for Wales (or VTW).

SUMMARY OF THE LAW OF RATING

1.9   Central to the very ancient system of rating law is the concept of the 'hereditament'. This is a technical term which really means no more than a unit of property. In general it will equate to the area which is occupied by one occupier. If the area occupied is not contiguous (if, for example, the occupier occupies several buildings in the same street, or several floors in an office block which do not directly intercommunicate), then it will normally be treated as more than one hereditament, unless one part is necessary for the effectual enjoyment of the other. Similarly, if one part of a contiguous area in one occupation is used for a wholly different purpose from the rest, it may be treated as a separate hereditament on that basis. There are a number of specific statutory provisions which apply to modify these tests in respect of certain sorts of property.

1.10 A hereditament cannot generate a liability to rates unless it is included on either the local or central list, maintained by Central Government. In order to be included in the list, it must be both 'relevant' and 'non-domestic'. Relevant hereditaments comprise lands, mines, advertising rights and rights to operate meters. Lands include buildings and anything above or below the surface of the earth, so most physical units of property will be caught by this definition. Non-domestic hereditaments are all those which are not exclusively made up of living accommodation or ancillary elements such as garages and outbuildings. This is again subject to specific statutory provisions covering certain classes of property (e.g. domestic renewable energy installations, short-stay accommodation, caravans and boats).

1.11 A hereditament included in the list will have a 'rateable value' attributed to it. The ratepayer will be liable to pay rates on a daily basis at a certain fraction of that rateable value. If the hereditament is occupied, then the occupier will be liable to pay rates. If it is unoccupied then, in general, the owner will be liable. Both of these terms, 'owner' and 'occupier', are subject to specific definitions. The owner does not mean the freehold owner, but the person entitled to possession of the hereditament. Occupation must be actual, beneficial, exclusive and not too transient if it is to be considered as rateable occupation. Each of these requirements is subject to extensive discussion in the case law. All of the concepts mentioned so far are discussed at length in Chapter 2, whereas the rules on liability are set out in Chapter 3.

1.12 The normal position as to liability may be modified by a number of exemptions, which remove property from the scope of rating altogether, and reliefs, which result in a reduction in the amount of liability. It is difficult to summarise the effect of these exemptions and reliefs shortly because they form a miscellaneous assortment which has no particular unifying principle. Exemptions relate, for example, to agricultural property, places of worship, parks and highways. There are also exemptions specific to unoccupied property, which encompass listed buildings, hereditaments of low value, and hereditaments which have only recently become unoccupied. Relief is available at the discretion of the billing authority in various specific cases, and there is also a general power to grant relief. All of the exemptions and reliefs are dealt with in Chapter 4.

1.13 The rating lists are maintained by Central Government in the person of the relevant Valuation Officer. Collection of rates is by and large dealt with by district councils, known in this context as 'billing authorities'. The collection procedure involves serving a demand notice or bill on each ratepayer. If the bill is not paid, further notices follow and proceedings can be taken in the civil courts or, in the vast majority of cases, in the Magistrates' Court. Successful proceedings in the Magistrates' Court lead to the granting of a liability order which can be enforced further in a number of ways. There are limited routes by which a liability order once made can be appealed or set aside. All of these matters are covered by Chapter 5.

1.14   A discrete function of billing authorities is the service of completion notices, which have the effect of deeming new buildings complete and therefore eligible to generate a liability to non-domestic rates. The relevant principles and the procedures for making and challenging these notices are outlined in Chapter 6. Chapter 7 deals with the means available to a ratepayer who wishes to recover rates from a billing authority when he believes he has made an overpayment.[1] The NDR Collection Regulations 1989 provide an automatic right to a refund where rates have been demanded and the amount paid is found to be in excess of the amount properly payable. There is also a right to sue the billing authority in restitution if, for various technical reasons, the right to a refund in the NDR Collection Regulations 1989 is not available.

SUMMARY OF THE LAW OF COUNCIL TAX

1.15   Council tax is payable in respect of 'chargeable dwellings'. A dwelling is effectively a domestic hereditament, and the extent of the dwelling will generally depend on the same principles as to the identification of hereditaments as apply in rating law. However, there are specific provisions which deal with aggregation (i.e. treating several dwellings as one dwelling) and disaggregation (i.e. treating one dwelling as several dwellings). The former are automatic when the criteria are satisfied, whereas the latter depend on the discretion of the listing officer, who maintains the list of chargeable dwellings. A dwelling will not be chargeable, and therefore will not generate any liability to council tax, if it is exempt. A large number of dwellings, categorised in Classes A to W, are exempt in this way. These provisions, which set the scope for the council tax, are dealt with in Chapter 8 (which also includes discussion of the completion notice regime for dwellings).

---

1 Throughout this text references to 'he/his' are not intended to be gender-specific and can equally be read as 'she/her'.

1.16   Chapter 9 deals with liability to council tax. There is a hierarchy of liability which sets out who will be liable in descending order. Liability will fall on the resident who has the strongest title to occupy the dwelling. If there are no residents, it will in normal circumstances fall on the owner. Regulations make provision in certain circumstances for the owner to be liable whether or not there are residents of the dwelling. 'Owner' and 'resident' are both given statutory definitions. A person is resident in a dwelling if he is 18 or older and 'has his sole or main residence in the dwelling'.

1.17   The amount of liability will depend on which valuation band the dwelling is in. These valuation bands are set by reference to capital values in 1991. The amount of liability will be modified in certain circumstances, however. The most common is council tax 'discount', which in general has the effect that dwellings with one resident benefit from a 25 per cent discount, and dwellings with no resident attract a 50 per cent discount (although these amounts can be altered by billing authorities in certain cases). In calculating the number of residents for these purposes, certain classes of persons are to be disregarded (e.g. students, care workers, patients in care homes).

1.18   An increasing move towards localism has seen the removal of council tax benefit, and in its place billing authorities are required to make a 'council tax reduction scheme' aimed at helping low-income council taxpayers. Similarly, billing authorities have powers to increase levels of council tax in respect of dwellings subject to forms of use or ownership apparently regarded as undesirable. They also have a general discretion to reduce council tax liability in specific cases or classes of cases.

1.19   The provisions on collection of council tax, set out in the CT Enforcement Regulations 1992, are in outline very similar to those in respect of rates. They are dealt with in detail, and relevant similarities and differences between the two regimes are pointed out, in Chapter 10, which also covers the provisions on obtaining repayments of council tax.

Part I

Non-domestic rating

2

Key concepts in rating law

2.1　There are a number of key concepts at the heart of rating law: the hereditament, the list, rateable value, the domestic/non-domestic distinction, ownership and rateable occupation. These concepts are dealt with in this chapter, and the rest of this work assumes knowledge of them. To a large extent they are interrelated. For example, the concept of rateable occupation is central to the definition of a hereditament.

THE HEREDITAMENT

2.2　Non-domestic rates are a tax on the occupation and ownership of property. Specifically, they are a tax on the occupation or ownership of units of property known as 'hereditaments'. The history of this ancient term is in the context of the buying and selling of property where it simply means something capable of being inherited. In the original context, the extent of the hereditament in question is defined by the document making the sale. In the rating context the concept is much more fluid as will be apparent from the rest of this section. Essentially the definition of a hereditament defines the unit of property on which tax is payable.

2.3　From the point of view of collecting rates, which forms the focus of this work, it is not normally necessary to debate the correct identification of the hereditament.[1] This is because for collection purposes hereditaments are conclusively defined with reference to the relevant list (on which, see below). It is however so fundamental to the operation of the system that a full treatment is given here.

1　Although see *Milton Keynes Council v Public Safety Charitable Trust* [2013] EWHC 1237 (Admin) at [49], where an argument about the extent of the hereditament defined in the list proved to be relevant in a collection context.

2.4   As it applies in the rating context, the term 'hereditament' is a statu-
       tory term with a statutory definition. However, the statutory definition
       of a hereditament is effectively circular and of no real help in identi-
       fying what is and is not a hereditament. LGFA 1988, s64(1) provides:

> An hereditament is anything which, by virtue of the definition
> of hereditament in s115(1) of the 1967 Act, would have been an
> hereditament for the purposes of that Act had this Act not been
> passed.

2.5   The General Rate Act 1967, s115(1) in turn provides as follows:

> 'hereditament' means property which is or may become liable to
> a rate, being a unit of such property which is, or would fall to be,
> shown as a separate item in the valuation list.

2.6   These provisions have been described in the Court of Appeal as 'leg-
       islative gobbledegook'.[2] This description is unkind but perhaps fair.
       The statutory provisions simply identify that the hereditament is 'a
       unit of . . . property'.[3] In practice, therefore, the principles as to what
       constitutes a hereditament are set out in decided cases.

2.7   These establish that the hereditament is equated with the unit of occu-
       pation.[4] Subject to certain qualifications, which are discussed below,
       the hereditament will be the area actually occupied as one unit. Except
       in cases of joint occupation of the whole,[5] there cannot be more than
       one rateable occupier of a particular hereditament. Therefore if a prop-
       erty is rateably occupied by more than one person it follows that it is

---

2 *Reeves (LO) v Northrop* [2013] EWCA Civ 362 at [9].

3 This requirement is already reflected in the common law's insistence that a hereditament must be capable of definition: *Peak (VO) v Burley Golf Club* [1960]
   1 WLR 568 at per Harman LJ p575–576.

4 This view of the law has been questioned by one member of the Court of Appeal
   in *Vtesse Networks Ltd v Bradford (VO)* [2006] EWCA Civ 1339 at [40] on the basis
   that 'until a hereditament is identified, occupation cannot arise, if for no other reason than that what amounts to occupation will ordinarily depend upon the physical
   and legal character of the hereditament'. The rest of the court did not endorse these
   observations so it remains correct to say that the hereditament is intrinsically linked
   to the concept of occupation.

5 As to which see para 2.113–119.

comprised of an equivalent number of hereditaments, whether or not the separate areas are 'structurally severed' from one another.[6]

Hereditament smaller than unit of occupation

2.8 There are qualifications on this principle which in appropriate cases will mean that the hereditament is smaller than the unit of occupation. The leading case on this is the Supreme Court's decision in *Woolway (VO) v Mazars LLP*.[7] It is important to note that the Supreme Court disapproved the Court of Appeal's earlier decision in *Gilbert (VO) v Hickinbottom & Sons Ltd*,[8] which for almost 60 years had been considered the leading case on the identification of hereditaments. Earlier authorities, which were decided in reliance on *Gilbert v Hickinbottom*, must accordingly be approached with care.

2.9 The *Mazars* case was about when floors in a modern office building would form separate hereditaments, but it also contains a definitive summary of the principles to be applied:

> First, the primary test is, as I have said, geographical. It is based on visual or cartographic unity. Contiguous spaces will normally possess this characteristic, but unity is not simply a question of contiguity . . . If adjoining houses in a terrace or vertically contiguous units in an office block do not intercommunicate and can be accessed only via other property (such as a public street or the common parts of the building) of which the common occupier is not in exclusive possession, this will be a strong indication that they are separate hereditaments. If direct communication were to be established, by piercing a door or a staircase, the occupier would usually be said to create a new and larger hereditament in place of the two which previously existed. Secondly, where in accordance with this principle two spaces are geographically distinct, a functional test may nevertheless enable them to be treated as a single hereditament, but only where the use of the one is necessary to the effectual enjoyment of the other. This last point may commonly be tested by asking whether the two sections could reasonably be let separately. Third, the question whether the use

---

6 *Allchurch v Hendon Union Assessment Committee* [1891] 2 QB 436 at p441–442.
7 [2015] UKSC 53.
8 [1956] 2 QB 40.

of one section is necessary to the effectual enjoyment of the other depends not on the business needs of the ratepayer but on the objectively ascertainable character of the subjects.[9]

2.10   On this basis, floors which were not adjacent and which did not inter-communicate directly were held to be separate hereditaments – they failed the primary, geographical test and were not necessarily enjoyed together. A majority of the judges also held that the same finding would apply even to adjacent floors which did not intercommunicate directly.[10]

Use for a wholly different purpose

2.11   The other common situation in which the hereditament does not cor-respond with the unit of occupation is where part of a larger unit is used for wholly different purposes from the rest. On this basis, a hotel and refreshment rooms forming part of a railway station were rated separately from the rest of the station.[11] The applicability of the princi-ple has been confirmed in modern times by the Court of Appeal[12] and by the Supreme Court in *Mazars*.[13]

2.12   It is to be noted that the principle has been said not to provide a 'con-clusive test', as the identification of the hereditament is a question of fact taking into account all material considerations.[14] Other consid-erations will perhaps be particularly important when this principle is applied to split a hereditament which is only part-occupied. This was done in *English, Scottish and Australian Bank v Dyer (VO)*,[15] in which a gardener's cottage (occupied) was treated as a separate hereditament to the mansion house (held in reserve for emergency use and con-sequently unoccupied). Another example is *British Railways Board v Hopkins (VO)*,[16] in which a part-occupied office block was treated as separate hereditaments.

 9   [2015] UKSC 53, per Lord Sumption at [12].
10   [2015] UKSC 53, per Lord Sumption at [21], Lord Gill at [43], Lord Neuberger at [56] and Lord Toulson at [63].
11   *North Eastern Railway Co v York Union* [1900] 1 QB 733.
12   *Coventry & Solihull Waste Disposal Co Ltd v Russell (VO)* [1998] RA 427.
13   [2015] UKSC 53, per Lord Sumption at [6]. See also Lord Gill at [39].
14   *Trafford MBC v Pollard (VO)* [2007] RA 49 at p63.
15   (1958) 4 RRC 27.
16   [1981] RA 328.

Cross boundary hereditaments

2.13   There is no longer any rule that a single hereditament crossing the boundary between rating districts should be treated as two separate hereditaments.[17]

Specific statutory provisions

2.14   Certain specific statutory provisions require property to be treated as a separate or combined hereditament.[18] In brief, these provisions apply to:

(1)   multiple caravan pitches on a caravan site;[19]
(2)   multiple moorings in the same ownership;[20]
(3)   property used for fuel production for electricity generation;[21]
(4)   docks and harbours;[22]
(5)   Crown property;[23]
(6)   telecommunications apparatus attached to a hereditament.[24]

THE LIST

2.15   Rating may be a tax on the occupation or ownership of hereditaments, but liability will not arise unless a hereditament is included in the lists maintained by the Valuation Office Agency. There is a 'local' list for each billing authority area,[25] and a 'central' list for certain prescribed hereditaments.[26] All 'relevant, non-domestic' hereditaments fall to be shown in the appropriate local list, unless they are required to be included in the central list.[27] These concepts are considered

---

17   Non-Domestic Rating (Miscellaneous Provisions) Regulations 1989, reg 6.
18   These provisions are made pursuant to LGFA 1988, s64(3)–(3A).
19   Non-Domestic Rating (Caravan Sites) Regulations 1990. If the units are not 'caravans' then the provisions will not apply: *Oades and Oades v Eke* [2004] RA 161.
20   Non-Domestic Rating (Multiple Moorings) Regulations 1992.
21   Non-Domestic Rating (Electricity Generators) Regulations 1991.
22   Non-Domestic Rating (Miscellaneous Provisions) (No 2) Regulations 1989.
23   LGFA 1988, s65A.
24   Non-Domestic Rating (Telecommunications Apparatus) (England) Regulations 2000, Non-Domestic Rating (Telecommunications Apparatus) (Wales) Regulations 2000.
25   LGFA 1988, s41–41A.
26   LGFA 1988, s52.
27   LGFA 1988, s42(1).

in turn below. The distinction between domestic and non-domestic hereditaments is of such importance that it is given its own section.

Relevant hereditaments

2.16   LGFA 1988 essentially defines a relevant hereditament as a hereditament consisting of property of the following descriptions: lands, mines, advertising rights and rights to operate meters.[28]

2.17   The category of 'lands' is by far the broadest and most common of the categories of 'relevant hereditament'. It encompasses buildings and anything above or below the surface of the earth.[29] At first sight the word 'lands' would seem to exclude chattels, but it has been confirmed by the House of Lords[30] that chattels can be rateable if they are enjoyed with land and enhance its value. Subsequent Court of Appeal decisions have confirmed this position, emphasising the need for a 'sufficient connection' with the land[31] and for a sufficient degree of permanence in the presence of the chattel.[32] Incorporeal rights (other than advertising rights and rights to operate meters, for which specific statutory provision is made) can be rateable in this category if the exercise of them involves the occupation of land.

2.18   'Mines' are split into two categories, namely 'coal mines' and 'mines of any other description, other than a mine for which the royalty or dues are for the time being wholly reserved in kind'.[33] These two categories in fact cover all mines other than mines in respect of which the royalties or dues are wholly reserved in kind. This category is therefore fairly comprehensive as far as mines are concerned.

2.19   The right to exhibit an advertisement on land is itself a relevant hereditament where that right is let out or reserved to someone other than the occupier of the land (or, where the land is unoccupied, to someone other than the owner of the land).[34] Similarly, a right to operate a gas

---

28 LGFA 1988, s64(4).
29 For example, telegraph wires in *Lancashire and Cheshire Telephone Co v Manchester Overseers* (1884) 14 QBD 267; a lake and the floating clubhouse upon it in *Thomas v Witney Aquatic Co Ltd* [1972] RA 493.
30 *LCC v Wilkins (VO)* [1957] AC 362, a case on builders' huts.
31 *Ryan Industrial Fuels Ltd v Morgan (VO)* [1965] 1 WLR 1347 at p1352.
32 *Field Place Caravan Park v Harding (VO)* [1966] 2 QB 484 at p498.
33 LGFA 1988, s64(4).
34 LGFA 1988, s64(2).

or electricity meter owned by someone other than the consumer is also a relevant hereditament. These provisions put beyond argument the rateability of what would otherwise potentially be contentious categories of hereditament.

Central list hereditaments

2.20   Some hereditaments do not fit neatly into the system of local lists corresponding to billing authority areas. This problem has been solved by the creation of a central list.[35] Regulations prescribe a list of designated persons, each of whom has an entry in the central list.[36] Every hereditament of a certain description owned or occupied by that person then forms part of one entry in the central list (sometimes known as a 'cumulo assessment'). The persons named are generally the operators of utility and communications businesses which naturally transcend billing authority areas.

NON-DOMESTIC HEREDITAMENTS

2.21   Non-domestic rates are commonly, but somewhat misleadingly, referred to as 'business rates'. In fact not just business or commercial hereditaments but all 'non-domestic' hereditaments fall to be included in the relevant list.

Domestic and composite hereditaments

2.22   Property is non-domestic if it is either composed entirely of property which is not domestic, or is a composite hereditament (i.e. partly domestic and partly non-domestic).[37] The result of this definition is that all hereditaments are to be included in the lists for non-domestic rating purposes unless they are wholly domestic. Where the hereditament is composite, however, its value in the list will be based purely on a non-domestic use of it.[38]

---

35 LGFA 1988, s52–53.
36 The current regulations are the (much amended) Central Rating List (England) Regulations 2005 and Central Rating List (Wales) Regulations 2005.
37 LGFA 1988, s64(8)–(9).
38 LGFA 1988, sch 6 para 2(1A).

Domestic property: the generic definition

2.23   A four part definition in LGFA 1988 sets out what is domestic property, as follows:

property is domestic if:

(a) it is used wholly for the purposes of living accommodation;
(b) it is a yard, garden, outhouse or other appurtenance belonging to or enjoyed with property falling within paragraph (a) above;
(c) it is a private garage which either has a floor area of 25 square metres or less or is used wholly or mainly for the accommodation of a private motor vehicle; or
(d) it is private storage premises used wholly or mainly for the storage of articles of domestic use.[39]

2.24   This definition, which applies in both England and Wales, is supplemented by certain specific provisions which are dealt with below. The third and fourth parts of this definition are self-explanatory and do not require any clarification, save to note that they do not overlap and therefore land or buildings used for car parking cannot fall within sub-paragraph (d) if they are not within sub-paragraph (c).[40] The first and particularly the second parts require some further discussion, however.

2.25   The expression 'used wholly for the purposes of living accommodation' in the first part of the definition needs only slight elaboration. On the one hand, it is not necessary for property to be used for all the purposes which could be described as living accommodation; it is thus possible for property to satisfy this part of the definition if it does not provide anywhere to sleep.[41] On the other hand, the purposes of living accommodation are not confined to the satisfaction of basic bodily needs, and accordingly if part of the property is in dedicated use for recreation or for working from home, this can still amount to use for the purposes of living accommodation.[42]

2.26   Although the word 'wholly' is used, this part of the definition incorporates the *de minimis* principle such that an insignificant amount of use for non-domestic purposes does not disqualify the hereditament

39 LGFA 1988, s66(1).
40 *Reeves (VO) v Tobias* [2010] UKUT 411148 (LC), [2011] RA 149 at [7]–[9].
41 *Lewis v Christchurch Borough Council* [1996] RA 229 at 233, in which beach huts were held to be domestic despite the absence of beds.
42 *Tully v Jorgensen* [2003] RA 233 at p240–241.

from being a 'domestic' one.[43] Nevertheless, rateability is likely to arise if the use is so significant that the accommodation is adapted so as to lose its domestic character.[44] Using property on a new housing estate as a dedicated 'show home' does not amount to use for the purposes of living accommodation.[45]

2.27 The second element of the definition is more convoluted. The terms 'yard, garden, outhouse' are straightforward terms with a commonly understood meaning. It is necessary to note, however, that they are all controlled by the phrase 'or other appurtenance'.[46] In other words, a yard, garden or outhouse must also be an 'appurtenance' if it is to fall within this provision. It is therefore necessary to consider the meaning of this somewhat archaic legal expression in more detail.

2.28 The term 'appurtenance' (like 'hereditament') originates in the context of buying and selling property. An appurtenance came to mean something that would automatically be transferred with the house, without it needing to be mentioned in the documents making the sale.[47] The practice has long since changed, however, such that it is now essential when selling property to refer to a plan. The term 'appurtenance' is therefore no longer one which is commonly understood or applied.

2.29 This problem has been resolved by reference to the concept of 'curtilage'. If property is within the curtilage of a house, then it would have passed automatically on a sale of the house and is therefore an appurtenance. It is accordingly not necessary, and indeed it may be misleading, to consider whether property would have passed automatically on a sale. Instead, the question is simply whether it falls within the curtilage of the house.[48]

---

43 *Fotheringham v Wood* [1995] RA 315 at p324.
44 *Tully v Jorgensen* at p241, illustrated by *Fotheringham v Wood* [1995] RA 315 (use of two rooms in a house as an accountant's offices, including for meetings) and *Bell v Rycroft* [2000] RA 103 (use of part of a house, including a specifically modified and extended garage, as a day nursery).
45 *Walker (VO) v Ideal Homes Central Ltd* [1995] RA 347.
46 *Martin v Hewitt (VO)* [2003] RA 275 at p279–280.
47 *Trim v Sturminster RDC* [1938] 2 KB 508, applied in the rating context by *Clymo v Shell-Mex & BP* [1963] RA 85.
48 *Methuen-Campbell v Walters* [1979] QB 525. See particularly the remarks of Buckley LJ at p542: the 'proposition, that because an item of property will pass *sub silentio* under such a conveyance of The Gables, it is therefore within the curtilage of The Gables, cannot in my opinion be maintained, for that confuses cause with effect'.

2.30   This may seem to substitute one problem for another, because the word 'curtilage' is no more capable of precise definition than is the word 'appurtenance'. However, there are a number of authorities explaining and applying the term 'curtilage' in a modern context, which are of some assistance. These confirm that the word 'curtilage' is not a term of art. Whether one property falls within the curtilage of another is a question of fact and degree, and the relevant considerations are: (1) the physical 'layout' of the two properties; (2) their ownership, past and present; (3) their use or function, past and present.[49] Where these considerations demonstrate an 'intimate association' between the two properties, or that they form an 'integral whole', then one will fall within the curtilage of the other.

2.31   In the context of the definition of domestic property, it is not necessary for the claimed appurtenance to be within the curtilage of one other hereditament. It can fall within the curtilage of several dwellings. Thus district heating systems[50] and a concierge room forming part of a residential development[51] have both been held to be domestic under this part of the definition. Where the living accommodation is divided between separate buildings which each have their own clearly defined curtilage, however, it will be more difficult to show that common facilities satisfy this provision.[52]

2.32   In most cases it will be very obvious what falls within the curtilage of a property used as living accommodation: namely the garden or yard and any outbuildings (such as stables). However, the principles above can be applied in cases of doubt to determine whether a particular property falls within the curtilage or not and therefore whether or not it is an 'appurtenance'.

2.33   There is a second element to this part of the definition. In order to qualify as domestic an appurtenance must be one 'belonging to or enjoyed with' the living accommodation. This requirement is satisfied where the two properties are in the same ownership, as the appurtenance will

---

49 *AG v Calderdale BC* (1983) 46 P&CR 399 as confirmed and applied in *Skerritts of Nottingham Ltd v Secretary of State (No. 1)* [2001] QB 59.

50 *Head (VO) v Tower Hamlets LBC* [2005] RA 177.

51 *The Collection (Management) Ltd v Walker* [2013] UKUT 0166 (LC), [2013] RA 311.

52 *Winchester City Council v Handcock (VO)* [2006] RA 265 – sewage works held not to be appurtenant to separate dwellings they served.

then 'belong to' the main living accommodation.[53] If the properties are not in the same ownership, then in order for one to be 'enjoyed with' the other it is probably necessary for them to be in common rateable occupation.[54]

2.34 Once it is established that property is within the curtilage of living accommodation (so as to be an appurtenance to it) and that it belongs to or is enjoyed with that accommodation, then it is domestic within this part of the definition. In contrast with the other three limbs of the definition, there is no requirement that the property be used in any particular way.[55]

2.35 It must be noted that the Upper Tribunal has recently suggested that a structure might fall within the curtilage of a dwelling and yet not be an appurtenance to that dwelling.[56] It is submitted that this approach is mistaken, being at variance with the approach indicated by the Court of Appeal, and that it leads to confusion by applying an unspecified further test beyond that set out above. The Tribunal gave the example of a structure located in the garden of a dwelling by a utility company under statutory powers – this, it was said, would form part of the curtilage but not be an appurtenance. On the contrary, it seems clear that such a structure would *not* form part of the curtilage – once it is remembered that considerations of ownership, use and function are relevant, as well as considerations of physical layout. If the law on 'curtilage' is applied correctly, there is no need to impose any further requirement.

2.36 In the same case, it was said to be common ground that the term 'other appurtenances' had to be construed according to the *eiusdem generis* rule.[57] In other words, that only an appurtenance coming within the same *genus* as a 'yard, garden or outhouse' would qualify. It is very doubtful whether this is correct, given that it is very difficult to describe any category which includes a 'yard, garden or outhouse' but excludes other 'appurtenances'. Furthermore, even if such a category could be

---

53 *The Collection (Management) Ltd v Jackson (VO)* [2013] UKUT 0166 (LC), [2013] RA 311 at [57].
54 *Head (VO) v Tower Hamlets LBC* [2005] RA 177 at [22].
55 *Levinson (VO) v Robeson and Gray* [2008] RA 60.
56 *The Collection (Management) Ltd v Jackson (VO)* [2013] UKUT 0166 (LC), [2013] RA 311 at [43].
57 Ibid. at [36].

discerned, there does not appear to be any good reason why the legislation should seek to make some appurtenances domestic but others non-domestic. Although this position was common ground in that particular case, therefore, it does not appear to be a good statement of the law.[58]

Domestic renewable energy

2.37   In England only, property is not domestic if the following conditions are met:

(1)   it is used wholly or mainly for the generation of electricity or the production of heat by a renewable means;[59]

(2)   it is situated in or on property which falls within the first two limbs of the generic definition set out above (i.e. living accommodation and appurtenances);

(3)   the majority of the electricity or heat is for people in the living accommodation; and

(4)   the capacity of the equipment does not exceed 10 kilowatts or 45 kilowatts thermal, as the case may be.[60]

2.38   This provision does not apply in Wales.

Short-stay accommodation

2.39   Certain property would satisfy the generic definition of domestic property, but is specifically excepted because of its commercial nature.

2.40   In England, short-stay accommodation is not domestic. This covers property which is wholly or mainly used in the course of a business of providing accommodation for short periods to individuals whose sole or main residence is elsewhere.[61]

2.41   There is an exception to the exception for what might be described as lodging houses or bed and breakfasts – in other words, properties where the intention of the owner or operator is to provide short-term

---

58   It is notable that only one of the parties was legally represented.

59   The eligible forms of generation are listed in the Climate Change and Sustainable Energy Act 2006, s26(2), and include biomass, photovoltaics, solar power and combined heat and power systems.

60   LGFA 1988, s66(1A)–(1B).

61   LGFA 1988, s66(2).

accommodation to no more than six people at once and as a subsidiary activity to the use of the property as his own home. These will remain domestic.[62] Where significant extra facilities have been provided for the benefit of the guests (such as en suite bathrooms and fire-protection measures), this will be an indication that the letting is not 'subsidiary' to the use of the property as a home.[63]

2.42 There is also a separate regime for self-contained self-catering accommodation provided commercially ('holiday cottage' and 'apart-hotel' type properties). Such accommodation is non-domestic where the intention of the freehold or leasehold owner is to make it available for short-term lettings totalling 140 days in the following year – unless it is the sole or main residence of anyone.[64] A drafting error in this part of the legislation means that the property will be domestic, and therefore not liable to non-domestic rates, when it is a self-contained part of a building that has only a freehold owner.[65]

2.43 Finally, timeshare accommodation is also deemed to be non-domestic.[66]

2.44 In Wales, the regime in respect of short-stay accommodation is the same save that for self-contained self-catering accommodation there are slightly more stringent requirements before this will be deemed to be non-domestic: it must have been made available for 140 days in the previous year and must have actually been used for 70 days.

Caravans and moorings

2.45 Caravan pitches and residential boat moorings are outside the scope of the generic definition of domestic property set out above. They will be domestic, together with any caravan or boat which may occupy them, if the caravan or boat in question is the 'sole of main residence of an individual'.[67] If so, then any 'any garden, yard, outhouse or other appurtenance belonging to or enjoyed with them' is also domestic property. This provision mirrors the wording of s66(1)(b) considered above and is highly likely to be construed in the same way.

---

62 LGFA 1988, s66(2A).
63 *Skott v Pepperell* [1995] RA 243.
64 LGFA 1988, s66(2)(b), (2B)–(2D).
65 *R (Curzon Berkeley Ltd) v Bliss (VO)* [2001] EWHC (Admin) 1130, [2002] RA 45 at [56]–[69].
66 LGFA 1988, s66(2E).
67 LGFA 1988, s66(3)–(4).

2.46 These requirements are more stringent than those for other property
because they incorporate a requirement that the boat or caravan be
the sole or main residence of an individual. The question of whether
a structure is indeed a 'caravan' may therefore be of some signifi-
cance. This question is to be decided in accordance with the relevant
legislation,[68] which provides that 'caravan' means:

> any structure designed or adapted for human habitation which
> is capable of being moved from one place to another (whether
> by being towed, or by being transported on a motor vehicle or
> trailer) and any motor vehicle so designed or adapted, but does not
> include—
>
> (a) any railway rolling stock which is for the time being on rails
>   forming part of a railway system, or
> (b) any tent.

2.47 In addition, specific provision is made for 'twin unit' caravans. The
fact that these cannot legally be moved by road is no obstacle to their
being held to be a caravan, as long as they do not exceed certain
dimensions.[69]

2.48 If a structure is not capable of being moved by road, therefore, or if
it exceeds the specified dimensions, it will not be a caravan and the
pitch it occupies will not be a 'pitch occupied by a caravan'. In those
circumstances the pitch/caravan will not be subject to the 'sole or
main residence' requirement, and the normal definition of domestic
property will apply. Similarly, even if the structure in question *is* a
caravan, the 'sole or main residence' requirement will not apply if it
is sufficiently permanent to be rateable with the land on the normal
principles applicable to chattels.[70] In such a situation, the property
being considered is not a 'pitch occupied by a caravan', but the pitch
and caravan together as one rateable hereditament. Thus the single
hereditament composed of pitch and caravan together is subject to the
generic definition of 'domestic' property, not the specific provisions on
caravan pitches.[71]

68 The Caravan Sites and Control of Development Act 1960, s29.
69 Caravan Sites Act 1968, s13.
70 These are discussed in para 2.17 and see specifically in relation to caravans *Field
   Place Caravan Park v Harding (VO)* [1966] 2 QB 484.
71 *Atkinson (VO) v Foster* [1996] RA 246.

Property not in use

2.49 Property not in use is domestic if it appears that, when next in use, it will be domestic.[72] What 'appears' to be the next use will be a question of evidence, and for reasons discussed elsewhere, that evidence will need to relate to the state of affairs on the day with reference to which the determination falls to be made.[73] In other words, it is probably not possible to consider matters with the benefit of hindsight.

RATEABLE VALUE

2.50 LGFA 1988 provides that the rateable value of a hereditament 'shall be taken to be an amount equal to the rent at which it is estimated the hereditament might reasonably be expected to let from year to year',[74] subject to certain assumptions as to the state of repair of the hereditament and the bearing of expenses. This beguilingly simple definition gives rise to a whole range of legal and practical issues which fall outside the scope of this book.

OWNERSHIP

2.51 In everyday use, the word 'ownership' is generally taken to refer to freehold or long leasehold ownership. However, it bears a particular meaning in the context of non-domestic rating, namely that the 'owner of a hereditament or land is the person entitled to possession of it'.[75]

2.52 There cannot in general be more than one 'owner' at a time,[76] and the phrase 'entitled to possession' means 'immediately entitled to possession'.[77] Thus a tenant under a lease which grants the right to exclusive possession of a hereditament will be the owner of that hereditament.

---

72 LGFA 1988, s66(5).
73 See discussion at para 4.140 (on the same expression when used in the context of charitable relief).
74 LGFA 1988, sch 6 para 2.
75 LGFA 1988, s65(1).
76 *Westminster City Council v Haymarket Publishing Ltd* [1981] 1 WLR 677.
77 *Sobam BV and Satelscoop BV v City of London Corporation* [1996] RA 93 at p108.

In general, it will therefore be straightforward to ascertain who the 'owner' of a particular hereditament is.

2.53 The cases also provide guidance in some less frequently encountered situations, as follows:

(1) Where an agent is entitled to possession, that entitlement will normally be attributed to his principal. For this reason a receiver appointed under a debenture which makes him the agent of the company in receivership will not be the 'owner' for rating purposes unless he has been given and has exercised a right to take possession on his own behalf;[78]

(2) A landlord who had taken proceedings to forfeit the lease was the owner where the tenant had accepted his repudiation of the lease by vacating the premises and returning the keys. The tenant would only remain the owner if he were disputing the forfeiture claim;[79]

(3) A trustee who is entitled to possession only in his capacity as a trustee will nevertheless be the owner.[80]

OCCUPATION

2.54 The concept of rateable occupation has been central to rating law since its beginnings in the seventeenth century. There is thus a large number of court decisions on the meaning of 'occupation' in this context. This case law continues to be relevant because the modern legislation provides that questions of occupation are to be determined in accordance with the rule which would have applied under the General Rate Act 1967.[81] Under that Act, the rules on occupation were those developed in the cases.

2.55 The effect of those cases is, in the modern law, universally acknowledged to be summarised in the proposition that four 'ingredients' must be present for rateable occupation to exist. Rateable occupation must be: (1) actual; (2) exclusive; (3) beneficial; and (4) non-transient. This formulation was first devised by counsel in the case of *John Laing & Son Ltd v Kingswood AAAC* and subsequently adopted by the

---

78 Ibid. at p107–108.
79 *Kingston upon Thames LBC v Marlow* [1996] RA 87 at p91.
80 *Marshall v Camden LBC* [1981] RVR 94 at p95.
81 LGFA 1988, s65(2).

Court of Appeal in that case.[82] It has subsequently been approved and relied upon in a vast number of rating cases, including implicitly by the House of Lords.[83]

2.56 The remainder of this section is devoted to a discussion of each of the ingredients. However, it is worth noting at the outset what will already perhaps be obvious – rateable occupation is not synonymous with legal possession, and the formal status of the occupier (i.e. whether he occupies under a lease, a licence or an easement) is immaterial.[84] This point is neatly illustrated by the fact that a squatter, who has no legal title at all, can be in rateable occupation.[85] That is not to say, however, that the substance of the legal title to occupy is always totally irrelevant, and those instances where it is relevant will be noted below.

2.57 Finally, it should be noted that the question of occupation is to be determined under LGFA 1988 on a daily basis. As such, the interpretation provisions explain that a hereditament is 'unoccupied or wholly or partly occupied on a particular day if (and only if) it is . . . unoccupied or wholly or partly occupied (as the case may be) immediately before the day ends'.[86] It is submitted that this provision does not require a focus on the situation at 11.59 p.m. The remarks of the Lands Tribunal in respect of a similar interpretation provision in LGFA 1988 are likely to apply here also:

> where s67(5) refers to the state of affairs existing immediately before the day ends it is not requiring that attention be confined to the particular activities being carried on at a precise moment in time. What has to be considered is the use of the property with all its features, and all that s67(5) does is to identify the material time by reference to which any change in the use of the property is to be related.[87]

---

82 [1949] 1 KB 344 at p350.

83 *LCC v Wilkins (VO)* [1957] AC 362.

84 *Westminster v Southern Railway* [1936] AC 511 at p529 and p533 per Lord Russell.

85 Modern examples of this are found in two Court of Appeal decisions: *Re: Briant Colour Printing Co Ltd (in liquidation)* [1977] 1 WLR 942 (factory workers engaged in a 'work in' in occupation) and *Westminster City Council v Tomlin* [1989] 1 WLR 1287 (unincorporated association of squatters known as 'The Guild of Transcultural Studies' in occupation of former embassy building). *Cory v Bristow* (1877) 2 App Cas 262 is a more venerable example.

86 LGFA 1988, s67(4).

87 *Tully v Jorgensen (VO)* [2003] RA 233 at p241–242.

Actual

2.58 In order for a hereditament to be occupied, there must be some actual use of it. As has been noted above, legal possession is not sufficient alone to amount to rateable occupation. When houses were still subject to rating, it was established on high authority that the owner of an empty house would not be rateable in respect of it.[88] Furthermore, it is not sufficient that the owner intends to occupy the hereditament at some point in the future. This was clearly established in the case of *Associated Cinema Properties v Hampstead*,[89] which concerned a house held as reserve offices. The Court of Appeal held that 'intention is relevant only when it goes to show a present occupation and user . . . It is never true to say that a mere intention to occupy in hypothetical circumstances which may never come into existence is equivalent to occupation.'[90] The occupation must be actual.

2.59 It is necessary to disregard certain items and activities when ascertaining whether there has been actual occupation.

2.60 First, it is necessary to disregard any 'plant, machinery or equipment' on the hereditament which either: (1) was used on the hereditament when it was last in use; or (2) is intended for use in or on the hereditament.[91] The correct approach is to ask whether there would be occupation *but for* the presence of the items in question.[92]

2.61 The term 'equipment' is quite broad on its face. The High Court has accepted, without any apparent unease, an agreement between parties in a case that it was broad enough to cover furniture.[93] It may be that the term is wide enough to cover anything intended for use on the premises. However, it is also possible that it would be construed according to the *eiusdem generis* rule with 'plant' and 'machinery'; i.e. so that it only covered equipment which was in the same category as plant and machinery. This would seem to include items designed for active use such as computers and furniture, but to exclude purely decorative items. The approach that the courts will take to this question remains unclear.

---

88 By remarks in two decisions of the House of Lords: *Liverpool Corporation v Chorley UAC* [1913] AC 197 at p211 and *Westminster City Council v Southern Railway Company* [1936] AC 511 at p529.

89 [1944] KB 49, and on appeal at [1944] KB 412.

90 [1944] KB 412 at p417.

91 LGFA 1988, s65(5). The predecessor to this provision was s46A of the General Rate Act 1967, inserted by the Rates Act 1984.

92 *Sheafbank v Sheffield MDC* [1988] RA 33.

93 Ibid. at p41–42.

2.62 Second, certain fairly specific uses connected with elections are also to be disregarded in deciding whether there is occupation.[94]

2.63 Items left on premises pending future usage can constitute actual occupation (as long as they do not fall to be disregarded under the provisions set out above). The best example is that from *Southend-on-Sea Corporation v White*[95] of a seaside shop which was only open in the tourist season. This case was decided before the statutory provisions on disregarding plant, machinery and equipment were introduced. The stock was removed in the winter, but the other items remaining on the premises were sufficient to amount to actual occupation. Such cases are to be contrasted with those where items of no value have been left behind simply because they are too much trouble to remove; the presence of such items may amount to actual occupation but will not suffice to be beneficial and therefore rateable.[96]

2.64 Slight acts of user can amount to actual occupation. Before agricultural hereditaments became exempt from rating, they supplied some of the clearest examples of slight use amounting to actual occupation.[97] Another striking case is *Liverpool Corporation v Chorley UAC*,[98] in which the use of over 1,000 acres of land as a gathering ground for a reservoir was held to amount to rateable occupation. It is notable, however, that even in that case the corporation had carried out various acts of occupation such as planting trees and also maintaining moorland so as to make it more suitable for the rearing and shooting of game. A more up-to-date example would be that of a public car park, in respect of which a local authority owner will be in rateable occupation even if it does not in fact exercise any control or levy any charges.[99]

---

94 LGFA 1988, s65(6)–(7).

95 (1900) 65 JP 7. See also *Gage v Wren* (1903) 67 JP 33, a case about a boarding house.

96 See at para 2.84 and onwards on beneficial occupation.

97 See, e.g., *R v Mirfield* (1808) 10 East 219: occupation of land by growth of saleable underwoods; *R v Heaton* (1856) 20 JP Jo 37: occupation of fields laid down to grass. Neither of these early authorities includes much discussion of the concept of rateable occupation.

98 [1913] AC 197.

99 *Slough Corporation v Lane* (VO) [1965] RA 24. By contrast, *City of London Real Property Co Ltd v Stewart* (VO) (1960) 6 RRC 398, [1962] RVR 246 suggests that a private owner will have to exercise some sort of control over the car park before he is regarded as the rateable occupier.

2.65   The modern law on slight usage as actual occupation is helpfully set
       out and summarised in the case of *Makro Properties Ltd v Nuneaton and
       Bedworth BC*.[100] The guiding principle was said to be as follows:

> The proper approach to be drawn from the authorities in my judg-
> ment is to consider both use and intention. If there is clear evidence
> or inference of an intention to occupy, such an intention taken
> together with the user, albeit slight, may be sufficient to amount
> to occupation as determined in *Melladew*. Slight user without such
> evidence of intention may not be sufficient.[101]

2.66   The implication of *Makro* is that exactly the same physical acts may
       amount to actual occupation or not depending purely on the intention
       of the 'occupier'. The question of intention to occupy will therefore
       need to be considered carefully in cases which are likely to be dis-
       puted. It is notable that, in *Makro*, the contention that the ratepayer
       intended to go into occupation was buttressed by the fact that it wished
       to reduce its liability to tax overall by incurring a short period of liabil-
       ity for occupied rates.[102] That is an incorrect approach which divorces
       the issue of 'intention' from its initial context; namely, whether the
       intention was to use the premises for the purposes of the ratepayer's
       business.[103] An intention to avoid paying rates is irrelevant to the use
       of the premises in the ratepayer's business. It is submitted that *Makro*
       was wrongly decided on this point.

2.67   Hereditaments used as warehouses may constitute an exception to the
       general rule that some physical use is required. This is because, as was
       observed in passing in the *Associated Cinema Properties* case,

> it is the business of a warehouseman to let unoccupied space for
> the purpose of the temporary storage of goods. To be available for
> this purpose, the warehouse must be empty, in whole or in part. If
> the owner makes a public announcement that he is prepared to let
> space in an empty warehouse, he may reasonably be said to be using
> that warehouse for the purposes of his business.[104]

---

100 [2012] EWHC 2250 (Admin).
101 Ibid. at [43].
102 Ibid. at [27], [44].
103 See *R v Melladew* [1907] 1 KB 192 at p203; *Hewson, Chapman & Co Ltd v Grimsby
     County Borough Rating Authority* (1953) 46 R&IT 703 at p704.
104 [1944] KB 412 at p415.

This doctrine goes back to the case of *R v Melladew*,[105] in which the only physical act of occupation was the display of a notice advertising space as available.

2.68 Even that limited physical act may not be necessary if the requisite intention is present, however. That appears to have been the finding in the case of *Hewson, Chapman & Co Ltd v Grimsby County Borough Rating Authority*,[106] which concerned timber sheds kept available for storage. There is no suggestion from the report that there was any physical act of occupation, but the Divisional Court nevertheless upheld a finding of rateable occupation.[107]

2.69 The relevance of this 'spare room' or 'reserved occupation' doctrine appears to be limited to warehouses and storage premises. Although it was applied to a factory in *Borwick v Southwark Corporation*,[108] the correctness of that decision has been doubted.[109] Attempts to apply it in other areas have not been successful.[110] Furthermore, it is necessary for the warehouse to be in the direct ownership or control of the warehouseman himself,[111] and the warehouseman must have the necessary intention to carry on his business from the warehouse during the relevant period.[112] It will be apparent from the qualifications in this paragraph that the 'spare room' doctrine will be relevant only relatively rarely, as an exception to the general rule that there must be some physical act of occupation.

105 [1907] 1 KB 192.

106 (1953) 46 R&IT 703.

107 The Court of Appeal's decisions in *Bexley Congregational Church Treasurer v Bexley LBC* [1972] 2 QB 222 and *Calmain Properties Ltd v Rotherham MBC* [1989] RA 305 both touch indirectly on the question of whether occupation of warehouses without any physical act at all is possible, without deciding the question.

108 [1909] 1 KB 78.

109 By the House of Lords in *Townley Mill Co (1919) Ltd v Oldham AC* [1937] AC 419 and also in *Associated Cinema Properties v Hampstead* [1944] KB 412.

110 See *Bexley Congregational Church Treasurer v Bexley LBC* [1972] 2 QB 222, where the Court of Appeal refused to apply the doctrine to a house kept available for occupation by a church minister.

111 In *Calmain Properties Ltd v Rotherham MBC* [1989] RA 305 the Court of Appeal refused to find occupation where the warehouse was owned by a property developer who was not likely to operate the warehouse himself, but through an intermediary.

112 *Overseers of Bootle v Liverpool Warehousing Company* (1901) 65 JP 740. The key distinction between this case and *Melladew* was the fact that there was no finding of intention.

Exclusive

2.70 In order to be rateable, occupation must be exclusive for the particular purpose of the occupier.[113] This means that the ratepayer (or ratepayers, in a case of joint occupation[114]) must have exclusive use of the hereditament he is occupying. Exclusive occupation is shown most easily by use plus an exclusive title to occupy, such as a lease.[115] This is because the existence of a lease will ordinarily show that no one else has a 'simultaneous right in respect of the same subject matter'.[116] However, if the occupation is in fact exclusive of all others then the form of the occupier's legal title does not matter.[117]

2.71 It follows that in the vast majority of cases this requirement will be easily and obviously satisfied. Where the ratepayer occupies a building then his occupation will naturally exclude all others from using the building. It will normally also be pursuant to a legal title which grants the exclusive right to use that building. The occupation will therefore generally be exclusive.

2.72 The requirement will usually only be contentious in two situations. The first is where the nature of the use is, for some reason, not such as to be obviously exclusive. The second is where there is more than one possible occupier.

2.73 'Right of way' type uses are an example of the first situation. A right of way type use is not naturally exclusive of all others. Where the proposed occupier is simply one of the several users, all with equivalent rights, then his occupation will accordingly not be exclusive. On this basis the occupation of wagon ways,[118] railway lines,[119] towing paths, rivers[120] and harbours have all been held not to amount to rateable occupation. Occupation of this nature can be exclusive and therefore rateable either

---

113 *John Laing & Son Ltd v Kingswood AAC* [1949] 1 KB 344 at p350.
114 See para 2.113 and onwards.
115 The form of the document is not important, but the substance is and should be analysed carefully. Does it in fact grant exclusive occupation? If so, this requirement is likely to be satisfied.
116 *Cory v Bristow* (1877) 2 App Cas 262 at p276.
117 *Westminster City Council v Southern Railway Co* [1936] AC 511 at p533.
118 *R v Jolliffe* (1787) 2 Term Rep 90. The Court's reasoning in this early decision appears to be based partly on the nature of the right as a licence or easement. It is apparent from later decisions that rateable occupation can occur even if the occupier has only a licence or easement.
119 *Burton-on-Trent CBC v Ind Coope Ltd* [1961] RVR 310.
120 *Doncaster UAC v Manchester Sheffield and Lincolnshire Railway Co* (1894) 71 LT 585, which approved earlier decisions to the same effect.

if the proposed occupier is in fact the only user of the right,[121] or if he is the owner of the land over which the right of way type use is exercised.[122]

2.74 Right of way type uses are simply one example of the cases in which points about the exclusive occupation can arise. However, the same issues can in theory arise in relation to any type of use, depending on the facts. To take one final example, the use of land as a golf course would normally amount to exclusive occupation. In *Peak (VO) v Burley Golf Club*[123] the golf courses were laid out on common land. Members of the public used the courses to play golf without paying the green fees charged by the golf club and without challenge. Various other uses made of the land by the public, such as picnicking, interfered with the clubs' use of the courses. As such, the Court of Appeal concluded that occupation was not exclusive.

2.75 The second situation in which this requirement is likely to be engaged is where there is more than one possible occupier of the hereditament. In these cases, it is necessary to decide whose occupation is 'paramount'. The leading case, which sets out the relevant principles to be applied, is *Westminster City Council v Southern Railway Co*,[124] in which the House of Lords considered whether various stalls and other structures within Victoria Railway Station were rateable as separate hereditaments; in other words, whether the occupation of the railway company or of the individual occupiers was paramount. Lord Russell set out the principles to be applied to such cases in the following paragraphs:

> The general principle applicable to the cases where persons occupy parts of a larger hereditament seems to be that if the owner of the hereditament (being also in occupation by himself or his servants) retains to himself general control over the occupied parts, the owner will be treated as being in rateable occupation; if he retains to himself no control, the occupiers of the various parts will be treated as in rateable occupation of those parts.[125]

---

121 *Burton-on-Trent CBC v Thomas (VO)* [1961] RVR 199 is an interesting contrast in this respect to the other railway case referred to earlier in this paragraph. The owner did use the railway line in this case but his use was *de minimis*.

122 See the nineteenth-century harbour cases in which harbour commissioners were frequently the owners, and therefore in exclusive occupation, of various works they had constructed but not the owners of the harbour itself: e.g. *New Shoreham Commissioners v Lancing* (1870) LR 5 QB 489.

123 [1960] 1 WLR 568.

124 [1936] AC 511.

125 [1936] AC 511 at p530.

> In truth the effect of the alleged control upon the question
> of rateable occupation must depend upon the facts in every
> case; and in my opinion in each case the degree of the control
> must be examined, and the examination must be directed to the
> extent to which its exercise would interfere with the enjoyment
> by the occupant of the premises in his possession for the pur-
> poses for which he occupies them, or would be inconsistent with
> his enjoyment of them to the substantial exclusion of all other
> persons.[126]

2.76 The *Southern Railway* case is also of some assistance in explaining
      how those principles are to be applied. The fact that the occupier is
      made subject to bye-laws and regulations did not make the owner's
      occupation paramount, nor did the fact that the owner controlled
      access to the extent of shutting the station precincts at night, thereby
      preventing access to the premises under consideration.

2.77 The clearest example of a 'subordinate' occupation is that of a lodger.
      This was referred to both in *Southern Railway* and in the earlier case
      of *Cory v Bristow*.[127] Although the lodger occupies his own room, the
      landlord also occupies the whole house for the purpose of his business
      of letting rooms. That occupation is paramount to the occupation of
      his lodgers because of the degree of control the landlord exercises –
      which will generally include control over access and the ability to
      enter the lodger's rooms if necessary.

2.78 Lodgers as such are of course no longer within the non-domestic rat-
      ing system. The House of Lords in *Southern Railway* did refer to the
      concept of a 'business lodger', who would not be in exclusive occupa-
      tion. Modern examples of the 'business lodger' arrangement are very
      few and far between, however. They seem confined mainly to cases
      where a contractor is granted permission to extract materials from
      land in other occupation.[128] Even in such cases, however, application

---

126 [1936] AC 511 at p532.
127 (1877) 2 App Cas 262 at p276.
128 See *Andrews v Hereford RDC* [1963] RA 75 (gravel pit operator with a
     non-exclusive licence and working subject to the direction of the owner held
     not to be in occupation). *Greenall (VO) v Castleford Brick Co* (1959) 5 RRC
     253 is another example of the extractor being held not to be in occupation,
     but the decision does not rely expressly on the issue of 'paramount control' in
     the landlord.

of the 'business lodger' principle is far from automatic.[129] An area in which the 'business lodger' principle may well be applied in future is in relation to self-storage facilities, which are very close in nature to a lodging house. There is an 'identity of interest' between the provider of the facility and the end users, in that one wishes to provide premises for storage and the other wishes to use such premises.[130]

2.79 It is sometimes suggested that two people can be in rateable occupation of the same subject matter but for different purposes. It is submitted that this view is incorrect. In fact, the requirement of exclusive occupation means that two people cannot be in rateable occupation of the same subject matter for different purposes.[131] Where two people are in 'occupation' of the same subject matter for different purposes, the correct approach is to decide whose occupation is paramount. Only the paramount occupation is rateable, as set out above. The Court of Appeal explained the concept in this way:

> I found some difficulty during the argument in understanding how the requirement that occupation must be exclusive could be reconciled with the well established rule that where there are two competing occupiers of the same hereditament, it is the paramount occupier who is rateable. If there are two occupiers of the same hereditament, how can either be exclusive? . . .
>
> Another way of explaining the difficulty might be that an occupier, in order to qualify for rateable occupation, has only to be in exclusive occupation for his own particular purposes. This does not exclude others from occupying the same hereditament for *their* particular purposes. Paramountcy is a way of choosing between exclusive occupiers in that sense.[132]

2.80 Various cases are commonly cited to show that two people can be in occupation of the same subject matter for different purposes. In

---

129 See the Court of Appeal's decisions in *Bartlett (VO) v Reservoir Aggregates* [1985] RA 191 and *Wimbourne DC v Brayne Construction Co Ltd* [1985] RA 234, a case on subcontractors.

130 See *Brook (VO) v Greggs* [1991] RA 61 and *Re the appeal of Heilbuth* [1999] RA 109. *Brook*, a case of market traders, suggests that where there is such an 'identity of interest' (which was not the case there), *Southern Railway* can be distinguished.

131 Two (or more) people may of course be in joint occupation for the same purpose: see para 2.113 and onwards.

132 *Wimbourne DC v Brayne Construction Co Ltd* [1985] RA 234 per Lloyd LJ at p243.

fact, these cases do not establish this proposition, as the following paragraphs show.

2.81 *Holywell Union v Halkyn District Mines Co*[133] is perhaps the authority most commonly cited in this connection. In this case the owner of land had granted a drainage company the exclusive right of drainage through a tunnel on his land, subject to the reservation to himself of mining rights. Pursuant to those mining rights a tramway had been laid along a part of the tunnel. Far from being a case of two people being held to be in rateable occupation, the drainage company was treated as being in exclusive occupation such that the owner was not in rateable occupation.[134] The only possible exception to this was the tramway, which the House of Lords did not find it necessary to consider.[135]

2.82 Cases involving hereditaments composed of wires (such as telephone or telegraph wires)[136] and mineral workings[137] are sometimes used to suggest that two people can be in occupation of the same subject matter for different purposes. These cases in fact simply concern the identification of different subject matter. Land is not limited to the surface but extends (in the words of the old maxim) 'up to the heavens and down to the inferno'.[138] It is therefore possible to have multiple hereditaments above and below each other just as it is to have hereditaments side by side.[139] These cases therefore simply relate to the definition of

---

133 [1895] AC 117.

134 See the judgments of Lord Herschell at p126, Lord MacNaghten at p131, Lord Davey at p134.

135 Lord Herschell at p126.

136 See for example, *Lancashire Telephone Co v Manchester Overseers* (1884) 14 QBD 267, *New St Helens and District Tramways Co v Prescot Union* (1904) 1 Konst Rat App 150 and *Jackson (VO) v London Rediffusion Service Ltd* (1951) 44 R&IT 439.

137 *Ryan Industrial Fuels v Morgan (VO)* [1965] 1 WLR 1347 is sometimes cited, but was actually a case where the occupier alone had extracted material from the spoil heap under consideration. The reservation of rights to the owner to extract certain materials did not stop the operator's occupation from being exclusive, and had the owner in fact extracted anything then the Court of Appeal held that the operator's occupation would still have been paramount and therefore exclusive.

138 For a more prosaic formulation in the rating context see *Westminster CC v Southern Railway* [1956] AC 511 at p529.

139 In *Electric Telegraph Co v Salford Overseers* (1885) 11 Exch 181, a 'wires' case, the examples were given of a building erected so that a street passes under it, or the upper floor of a building in multiple occupations.

separate but adjoining areas in different occupations,[140] as do cases on growing crops.[141] Had the tramway in the *Holywell* case been considered a separate hereditament it would also have been explicable on this principle.

2.83 Finally, it has sometimes been suggested that there is a presumption that the owner is also the occupier. The correct view is that there is no such presumption. Instead, ownership is simply one circumstance to be taken into account. If there is evidence of use, but it is not clear who is using the hereditament, it may be inferred that the use is by the owner. But that inference can be, and in practical terms almost always will be, displaced by evidence. This is the extent of any 'presumption'.[142]

Beneficial

2.84 Beneficial means valuable, not profitable.[143] Therefore if an occupier is in occupation in order to discharge a statutory duty, exercise a statutory power[144] or fulfil a philanthropic purpose[145] his occupation will still be considered beneficial and therefore rateable, even if the enterprise runs at a loss.

2.85 There is no requirement for profit in order to find beneficial occupation. Accordingly, an occupier who is making use of a hereditament will still be in beneficial occupation even if he does not maximise the benefit,

---

140 *Bartlett (VO) v Reservoir Aggregates* [1985] RA 191 was a case about extraction of minerals from the bottom of a reservoir, which were held to be in separate occupation to the reservoir itself. The Court of Appeal at p200 referred to the reservoir as 'the adjoining hereditament'; i.e. adjoining the mineral workings.

141 Such as *R v Mayor of London* (1790) 4 TR 21 and *Back v Daniels* [1925] 1 KB 526. Reference back to common law decisions in relation to trespass, such as *Cox v Glue* (1848) 5 CB 533 (cited in *Back v Daniels*) suggests these cases are about occupation of different and adjoining strata of the land.

142 *Re Briant Colour Printing Co Ltd (in liquidation)* [1977] 1 WLR 942 per Buckley LJ.

143 This was established in a series of decisions of the House of Lords starting in *Jones v Mersey Docks and Harbour Board* (1865) 11 HL Cas 443 and running through *LCC v Erith Churchwardens* [1893] AC 562 to *West Kent Main Sewerage Board v Dartford UAC* [1911] AC 171.

144 As, for example, in *Cumbria CC v Sture (VO)* [1974] RA 472, which concerned a nature reserve.

145 *Bowes Museum and Park Trustees v Cutts (VO)* (1950) 43 R&IT 881 at p908–909.

or chooses not to derive a financial benefit from the occupation.[146] Such cases are sometimes described as cases of 'potential beneficial occupation',[147] but this is somewhat misleading.[148] The benefit is real, but there is a potential profit which the occupier has not taken. As profit is not required in order to find beneficial occupation, the fact that the occupier chooses not to profit by his occupation is irrelevant.

2.86   Land may be incapable of beneficial occupation by anyone. Such land is sometimes said to be 'struck with sterility'.[149] This may be a result of the natural condition of the land – if it is a 'barren rock' with no value to anyone.[150] Or it may be because the land is dedicated to public use such that no one person can occupy it. Public highways and public parks are the best examples of the latter category.[151] In other cases, whether there is beneficial occupation will depend on what is actually done on the hereditament.

2.87   In order for occupation to be beneficial, the value to the occupier must be derived from some use of the hereditament according to its nature. In *Arbukcle Smith v Greenock Corporation*[152] the House of Lords considered the case of a warehouse being converted into a bonded store. These acts of use were not sufficient to give rise to rateable occupation because, although beneficial in one sense, they did not amount to 'use of the premises according to their nature'[153] nor to 'enjoyment of the value of the building as a warehouse'.[154]

---

146   *R v Heaton* (1856) 20 JP 37; *Kingston-upon-Hull Corporation v Clayton* [1963] AC 28 per Lord Guest. For an illustration, see the following three cases about car parks, in all of which the occupiers did not charge but were held to be in beneficial occupation: *City of London Real Property Co Ltd v Stewart (VO)* (1960) 6 RRC 398, *Slough Borough Council v Lane (VO)* (1965) 11 RRC 43, *Wokingham Corporation v Walker (VO)* (1965) 11 RRC 64.

147   See, for example, *Winstanley v North Manchester Overseers* [1910] AC 7 at p15.

148   As Widgery J appears to have acknowledged in *Liverpool Coporation v Huyton-with-Roby UDC* (1964) 10 RRC 256 at p260–261.

149   *London County Council v Erith Overseers* [1893] AC 562 at p591.

150   The expression 'barren rock' was used by Lord Cranworth in *Jones v Mersey Docks* (1865) 11 HL Cas 443 at p507.

151   *Lambeth Overseers v LCC* [1897] AC 625. For the modern law on the exemption enjoyed by public parks, see para 4.50 and following.

152   [1960] AC 813.

153   Per Viscount Kilmuir at p821.

154   Per Lord Radcliffe at p829.

2.88   For the same reasons, it is well established that having a caretaker in occupation is not sufficient, unless the caretaker is also there to perform other functions of benefit to the owner.[155] Putting up a 'to let' board is not sufficient to constitute beneficial occupation of the premises either.[156]

2.89   There is a fine line between occupation which is beneficial and occupation which is not beneficial. A good example of this is in the authorities dealing with items left behind when an occupier has otherwise vacated the premises. The principle was established in *London County Council v Hackney Borough Council*.[157] In that case two cupboards and an old mangle had been left on premises. The court held that they did not amount to beneficial occupation; they 'were kept on the premises simply because they were of no practical importance and though they had to be removed before any valuable use could be had of the premises, were not worth the trouble of removing for their own sake'. The principle has been approved and applied in later cases.[158] An example of a decision the other way is *Appleton v Westminster Corporation*,[159] in which books were left in a shop after it had closed. They appear from the report not to have been particularly valuable, but they did not fall into the category of 'rubbish' or 'valueless chattels' and so did give rise to beneficial occupation.

---

155   An early decision to this effect is *R v Morgan* (1834) 2 Ad & El 618n. Fuller reasoning is found in *Bertie v Walthamstow* (1904) 68 JP at 545. *Hicks v Dunstable* (1883) 48 JP 326, which appears to suggest otherwise, is probably wrongly decided. The caretaker would not be in occupation either because he was there as the servant of the owner: *Yates v Chorlton-upon-Medlock Union* (1883) 47 JP 630.

156   *Smith v New Forest Union* (1890) 54 JP 324 and see the fuller statement of the facts at (1889) 53 JP 661. Note that this was a case where the whole hereditament was to let. It is different in that respect from the cases discussed at para 2.67, where a warehouseman offers space in his warehouse to let as part of his business.

157   [1928] 2 KB 588 at p598.

158   *Offaly CC v Williams* [1930] IR, *Ministry of Transport v Holland* (1963) 14 P& CR 259, [1962] RVR 552 (Court of Appeal), *Holyoak (VO) v Shepherd* [1969] RA 524.

159   [1963] RA 169.

2.90   If the value derived by the occupier from his occupation is very slight, then arguably beneficial occupation will not arise. The principle of *de minimis non curat lex* (the law is not concerned with trivial matters) can be applied to disregard a very slight benefit. The Magistrates' Court in *Wirral BC v Lane* took this approach where the occupier had stored valuable furniture on the premises and it was upheld on appeal to the High Court.[160] In the recent *Makro* case[161] this decision was discussed in the context of 'actual' occupation. In fact the findings, and the application of the *de minimis* maxim, make more sense in the context of beneficial occupation.

2.91   Neither of the two recent decisions of the High Court on slight occupation (*Makro*[162] and *Sunderland v Stirling*[163]) has considered the application of the *de minimis* maxim to beneficial occupation.

2.92   *Makro*[164] concerned the storage of pallets of paperwork for a limited period as a rates-avoidance measure. The High Court found a benefit to the occupier because the documents were required by law to be retained and so had to be stored somewhere. Given that the documents had been brought to the premises under consideration specifically as part of the rates-avoidance scheme, at some cost, it would seem that the benefit deriving from the storage of them at those premises was fairly minimal. However, the court did not consider the application of the *de minimis* maxim, and found beneficial occupation.[165]

2.93   *Sunderland v Stirling*[166] concerned a piece of broadcasting equipment measuring around $100 \times 100 \times 50$mm. Unsurprisingly, the focus of the *de minimis* maxim was again on the physical extent of the use. It was not considered in relation to the requirement of beneficial occupation, in respect of which the arguments were rather different.

160 [1979] RA 261. A very early example of the application of the maxim to beneficial occupation is arguably found in *R v Morgan* (1834) 2 Ad & El 618n (in which a 'trifling exception' is disregarded), although the report is not particularly clear.

161 *Makro Properties Ltd v Nuneaton and Bedworth BC* [2012] EWHC 2250 (Admin) at [41]–[43].

162 [2012] EWHC 2250 (Admin).

163 [2013] EWHC 1413 (Admin).

164 [2012] EWHC 2250 (Admin).

165 It is not clear whether the court also considered that the tax-avoidance advantage of incurring liability to occupied rates was a 'benefit' to the occupier (at [46]). Such a benefit is totally divorced from use of the premises according to their nature, so it is hard to see how it could support a finding of beneficial occupation.

166 [2013] EWHC 1413 (Admin).

2.94 As such, there is no reason why the *de minimis* maxim should not be considered and applied in relevant cases to the requirement of beneficial occupation. A slight benefit which falls below this threshold will not be sufficient to make occupation rateable.

Non-transient

2.95 In order to be rateable, occupation must be permanent or non-transient. The requirement of permanence, or non-transience, is firmly established as one of the four ingredients of rateable occupation. It was listed as one of the four ingredients of rateable occupation in *John Laing*,[167] and therefore forms part of the summary of rateable occupation that has been approved and applied many times:[168] occupation 'must not be for too transient a period'. Even before this, however, the House of Lords in *Westminster v Southern Railway* spoke of the need for rateable occupation to have 'some degree of permanence: a mere temporary holding of land will not constitute rateable occupation'[169] and in *Cory v Bristow* of 'a permanent and profitable occupation of land'.[170]

2.96 The central consideration is the character of the use. It must, in the almost poetic words of one of the early cases, be the occupation of a settler not a wayfarer.[171] This test was recently elaborated upon by the Court of Appeal:

> The occupier must have put down some roots which tie him to indefinite occupation and make him a settler in the property rather than a wayfarer passing by. The settler will have adopted the property for his residence for settled purposes as part of the regular order of his life for the time being. The wayfarer is an itinerant of no fixed abode.[172]

---

167 [1949] 1 KB 344 at p350.
168 Including by the House of Lords in *LCC v Wilkins (VO)* [1957] AC 362. Most recently, see the Court of Appeal's decision in *Reeves (VO) v Northrop* [2013] 1 WLR 2867, [2013] EWCA Civ 362.
169 [1936] AC 511 at p529.
170 (1877) 2 App Cas 262 at p276, citing *Forrest v The Overseers of Greenwich*, (1858) 8 E&B 890 at 900.
171 *R v St Pancras AC* (1877) 2 QBD 581 per Lush J at p589.
172 *Reeves (VO) v Northrop* [2013] EWCA Civ 362, [2013] 1 WLR 2867.

2.97   In assessing that character, 'permanence' is not to be taken literally. An occupation can be of fixed duration and still be 'permanent'. Similarly, an occupation which may be brought to an end by the giving of notice can also be 'permanent'. This was explained by Lord Radcliffe in *LCC v Wilkins* as follows:

> Certainly it is true that the law demands that an occupation to be rateable should be permanent. But then it is equally certain that permanence does not connote what it might appear to in this connexion. It is rather easier to say what it does not mean than what it does. An occupation is not the less permanent because it is that of a lessee who holds under a lease for a fixed term. In other words, there is permanent occupation however clearly the end may be in sight. More than that, an occupation can be permanent even though the structure or other chattel which is the means of occupation is removable on notice.[173]

2.98   The duration of the occupation is a relevant factor, but is not the only consideration.[174] At one stage the cases suggested that there was a 'rule of thumb' that, for temporary structures, they must be present for 12 months or more before becoming sufficiently permanent to be rateable.[175] An attempt to apply this 'rule' to mineral excavations was firmly rejected by the Court of Appeal in *Dick Hampton (Earth Moving) Ltd v Lewis (VO)*.[176] The court acknowledged that the length of the occupation was one factor, but held that the character of the use was equally or more important in determining whether rateable occupation existed. To this extent, the words 'permanence' and 'transience' had become terms of art and had lost their ordinary meaning.[177] The massive and inherently permanent nature of the mineral extraction in that case meant that rateable occupation existed even though the operations had only carried on for around seven months.

---

173   [1957] AC 362 per Lord Radcliffe at p381.
174   *Reeves (VO) v Northrop* [2013] 1 WLR 2867, [2013] EWCA Civ 362 at [25]–[26].
175   Articulated by the Lands Tribunal in *Robert McAlpine v Payne* [1969] RA 368.
176   [1976] QB 254.
177   See per Ormrod LJ at p268. At p261 Lord Denning appears to have doubted whether there was a requirement of permanence at all, except in relation to temporary structures.

2.99   The intentions of the occupier are also relevant.[178] If the occupier intends to continue his occupation indefinitely, then that counts in support of a finding of permanence.[179] Where there are no arrangements for a use or occupation to continue for the long term, then that does not assist the occupier in attempting to show permanence of occupation.[180] In this context, the nature of the agreement or legal right under which the use takes place may be of some relevance. If the use is pursuant to a long-term agreement, that tends to show permanence.[181] It is probably going too far to suggest that any use pursuant to a lease or a freehold interest will be 'permanent'[182]; certainly there are examples from the cases where such uses have been held to be too transient to be rateable.[183]

2.100   Finally, it is relevant to have regard to the nature of the thing being occupied, in particular, whether the rateable occupation is said to arise by the presence of a temporary structure. Such structures are by nature more likely to be a 'wayfarer'. A more stringent approach to permanence is therefore required compared to those cases which concern permanent hereditaments such as land or buildings.

2.101   Three misconceptions about the requirement of non-transience or permanence are dealt with below.

2.102   It is sometimes assumed that any occupation of a permanent building will satisfy this requirement. This idea stems from a misreading of the Lands Tribunal's remarks in *Sir Robert McAlpine*: 'for permanent buildings there is no fixed period: theoretically at any rate occupation of a house or a shop for even a day could attract rateability'.[184]

---

178   *Cory v Bristow* (1877) 2 App Cas 262 per Lord Hatherley at p279 is an early example of high authority in which intention was relevant; evidenced in that case by the permanent nature of moorings in a river bed.

179   *Renore Ltd v Hounslow LBC* [1970] RA 567 at p574.

180   *West Yorkshire MCC v Miller (VO)* [1984] RA 65 at p71.

181   See in this respect the contrasting decisions on transience (reached in respect of the same amusement fair operator) in *Hall (VO) v Darwen BC* (1958) 51 R&IT 9 at p11 and *Liverpool Corporation v Huyton-with-Roby Urban District Council* (1964) 10 RRC 256. Later cases have held that the relevant point of distinction was the existence of a seven-year agreement in the first case: see *Moore v Williamson (VO)* [1973] RA 172 at p178.

182   As Ormrod LJ did in *Dick Hampton v Lewis* [1976] QB 254 at p268.

183   See, for example, *West Yorkshire MCC v Miller (VO)* [1984] RA 65.

184   [1969] RA 368 at p376. *West Yorkshire MCC v Miller (VO)* [1984] RA 65 at p70 also treated land as having 'a permanent quality, unlike building huts'.

These remarks are not binding authority (as the Tribunal was not considering permanent buildings in that case) and must be treated with great caution following the Court of Appeal's decision in *Dick Hampton*, discussed above. Furthermore, the Tribunal did not say that one day of occupation always or generally *would* be rateable, just that occupation for one day 'theoretically . . . could' amount to rateable occupation. Whether or not a short period of occupation of a building will be rateable will depend on the application of all the factors set out above.

2.103   'Permanent' does not mean the same as 'continuous' or the opposite of 'intermittent'. The mistaken view that this is the meaning of permanence has featured for some considerable time in the textbook *Ryde on Rating*. It was mentioned, but not endorsed, by the House of Lords in *LCC v Wilkins*.[185] However, this view is inconsistent with the authorities: first, because it was specifically rejected in an early case;[186] second, because there are several cases of intermittent occupation which was nevertheless held to be sufficiently permanent;[187] third, because it is inconsistent with the fact that the duration of the occupation is an important and relevant consideration in identifying permanence.[188]

2.104   The requirement of permanence has not been altered by LGFA 1988 or by the introduction of an exemption from empty rates following six weeks of occupation.[189] It is true that some of the early cases do justify the requirement for permanence by reference to the nature of the rate as an annual charge. Following LGFA 1988, it is a daily charge,[190] and the regulations made under LGFA 1988 clearly

---

185 [1957] AC 362 per Lord Radcliffe at p381.

186 *R v St Pancras Assessment Committee* (1877) 2 QBD 581 per Lush J at p589: 'I do not agree with Mr. Castle that the word "permanence" is used in this class of cases in the sense of being continuous as to its use'.

187 Those are the cases dealing with markets which are held only on certain days of the week. In *Williams v Wednesbury and East Bromwich Churchwardens* (1886–1890) Ryde's Rating Appeals 327 at 330, the 'intermittent' point was specifically argued and was rejected.

188 As was acknowledged by the House of Lords in *LCC v Wilkins* [1957] AC 362 and the Court of Appeal in *Dick Hampton v Lewis* [1976] QB 254, per Roskill LJ at p265, Ormrod LJ at p267–268.

189 In The Non-Domestic Rating (Unoccupied Property) (England) Regulations 2008, and see para 3.23.

190 See Chapter 3.

contemplate the possibility of occupation for a six-week period. However, LGFA 1988 itself specifically preserves the old rules on rateable occupation.[191] As such it has not introduced any change to the requirement of permanence.

OCCUPATION: SPECIAL CATEGORIES

2.105 Certain situations are subject to specific legal principles or statutory provisions concerning occupation.

Occupation by employees

2.106 The occupation of an employee may sometimes be attributed to his employer. This will in fact often be the case because an employer generally does not occupy except through the presence of his employees. Disputed cases have mainly arisen where the question is whether domestic premises are also to be treated as occupied by the employer.

2.107 There are numerous examples of such cases. The principles, however, are set out in two decisions of the House of Lords. The first, *Glasgow Corporation v Johnstone*[192] concerned the occupation of a house by a church officer adjoining the church. The principles to be applied were expressed as follows by Lord Hodson (using the now old-fashioned terms 'master' and 'servant' rather than 'employer' and 'employee'):

> The distinction is usually shortly stated in this way: if the servant is given the privilege of residing in the house of the master as part of his emoluments the occupation is that of the servant. He is treated for occupation purposes as being in the same position as that of a tenant. If, on the other hand, the servant is genuinely obliged by his master for the purposes of his master's business or if it is necessary for the servant to reside in the house for the performance of his services the occupation will be that of the master.[193]

2.108 This statement of principle was further clarified in *Northern Ireland Commissioner of Valuation v Fermanagh Board of Education*,[194] a case

191 LGFA 1988, s65(2).
192 [1965] AC 609.
193 [1965] AC 609 at p626.
194 [1969] 1 WLR 1708.

about school teachers. The House of Lords explained that the key question was whether the requirement to live in the house formed part of the contract of employment.[195] It could form part of that contract either expressly or by implication. Lord Upjohn summarised the principles as follows:

> First, if it is essential to the performance of the duties of the occupying servant that he should occupy the particular house or it may be a house within a closely defined perimeter, then, it being established that this is the mutual understanding of the master and the servant, the occupation for rating and other ancillary purposes is that of the master and not of the servant. In truth and in fact, in such a case, if the necessity for occupation is not expressed in the contract between master and servant it must, of course, be an implied term thereof in order to give efficacy to the contract between the master and the servant. This is obvious, although it seems never to have been pointed out before, perhaps because it is so obvious. Secondly, there is the case where it is not essential for the servant to occupy a particular house or to live within a particular perimeter, but by doing so he can better perform his duties as servant to a material degree; then, in such case, if there is an express term in the contract between master and servant that he shall so reside, the occupation for rating and ancillary purposes is treated as the occupation of the master and not of the servant.[196]

2.109   It was suggested that these principles would be applicable only to dwelling houses occupied as part of an employer–employee relationship.[197] They have, however, been applied in other analogous contexts: namely, to members of a religious order[198] and to a congregational minister (who may or may not have been an employee).[199]

---

195 This is particularly stressed by Lord Diplock. Lord Pearson considered that the question 'should depend on the dominant character and purpose of the occupation rather than the contractual arrangements', p1725.
196 [1969] 1 WLR 1708 at p1722.
197 Lord Diplock cautioned against extending what he regarded as an 'anomalous' situation: [1969] 1 WLR 1708 at p1732.
198 *Commissioner of Valuation v Redemptorist Order* [1972] RA 145.
199 *Belmont Presbyterian Church v Commissioner of Valuation* [1971] RA 369.

Those applications of the principles are derived from older case law which was cited with approval by the House of Lords; apart from such situations it seems unlikely that the principles expressed by the House of Lords can be applied unless there is an employee/employer relationship.

Occupation by agents

2.110   The fact that an occupier is an 'agent' of another party does not prevent him from being in occupation. Agency is only relevant insofar as it shows a degree of control similar to that in an employee/employer relationship; i.e. a day-to-day control over the activities to be carried out on the hereditament. These points are clearly established by the decision of the House of Lords in *Solihull Corporation v Gas Council Corporation*.[200] Lord Reid held as follows:

> 'Agent' is sometimes rather loosely used to denote a person who, though not a servant, has obliged himself to accept a measure of control comparable with the control exercisable by a master over a servant – a person who is merely the hand of his 'principal' . . . I do not doubt that an owner of property could so subject himself to the control of another that that other person could be held to be the occupier, although he never was present on the property and exercised his control solely by giving orders to the owner.[201]

2.111   However, that was not the situation in the *Gas Council* case and there is no reported case where this has been found to have occurred.[202] It may seem unlikely that such a degree of control would occur outside of an employee/employer relationship, although it is not impossible, particularly where the two parties are corporate bodies and one is a wholly owned subsidiary of the other. The discussion on occupation by receivers, below, is also relevant.

200 (1962) 9 RRC 128.
201 (1962) 9 RRC 128 at p132–133.
202 A similar argument was advanced in *Greenwoods Building Industries Ltd v Sainsbury and Hatton (VO)* (1955) 48 R&IT 155, and rejected on the basis that the 'servant' company was in fact merely a subcontractor.

Occupation by receivers

2.112   A receiver will generally not be in occupation, because he is likely to be present as an agent of some other party. In many cases he will be there as an agent of the company whose business he is managing.[203] Even where a receiver is not an agent of the company,[204] however, or where the company has gone into liquidation such that he is no longer its agent,[205] it is still unlikely that a receiver will be held to be in rateable occupation. In order for that to happen, it must be shown that he has put the company out of occupation and/or taken occupation independently. This will rarely happen.

Joint occupation

2.113   It is well established that two or more occupiers can jointly occupy a whole hereditament. This was confirmed in the House of Lords by Lord Diplock as follows:

> Parliament cannot have intended to impose separate and independent liabilities to pay the rates for the same hereditament upon more than one person except where their legal right of occupation is a joint right, as in the case of joint tenants. In English law, therefore, although there may be a joint occupation of a single hereditament, there cannot be rateable occupation by more than one occupier whose use of the premises is made under separate and several legal (or equitable) rights.[206]

2.114   The possibility of joint occupation is also recognised in LGFA 1988, which provides that regulations can be made to deal with cases of joint occupation.[207] These are found in Part II of the Non-Domestic

---

203   Receivers were found not to be in occupation on this basis in *Ratford v Northavon DC* [1987] QB 357.

204   As in the early cases of *In re Marriage, Neave & Co* [1896] 2 Ch. 663, *National Provincial Bank of England v United Electric Theatres* (1915) 85 L.J. Ch. 106 and *Cyton v Palmour* [1945] K.B. 426.

205   As in *Boston BC v Rees* [2001] EWCA Civ 1934, [2002] 1 WLR 1304.

206   *NICV v Fermanagh Protestant Board of Education* [1969] 1 W.L.R. 1708 at p1728 per Lord Diplock.

207   LGFA 1988, s50.

Rating (Collection and Enforcement) (Miscellaneous Provisions) Regulations 1990. Regulation 3 provides as follows:

(1) This regulation applies in any case where (apart from this regulation) there would at a particular time be more than one occupier of a hereditament which is shown in a local non-domestic rating list, or of part of such a hereditament, or more than one owner of the whole of an unoccupied hereditament so shown.

(2) Where this regulation applies— . . .

    (b) as regards any time in a chargeable financial year when there is more than one such owner or occupier, the owners or occupiers shall be jointly and severally liable to pay the amount that would have been payable by way of non-domestic rate with respect to that time if there were only one such owner or occupier (as the case may be).

2.115 These Regulations also provide a default mechanism whereby the occupiers can recover the cost of paying the rates from each other, in case no other such mechanism exists.[208]

2.116 The most obvious example of joint occupation is that of partners in a firm. Each is in occupation of a hereditament occupied by the firm. Another example would be of owners in possession. Property can be owned either as 'joint tenants' or as 'tenants-in-common'. Joint tenants each own the whole of the property. Tenants-in-common each own a defined share of the property. Lord Diplock, in the passage cited above,[209] appears to suggest that only joint tenants can be in joint occupation. There is a certain logic to this, as tenants-in-common may hold widely divergent shares in the property so it seems unfair to make each of them liable in full for the rates due. Nevertheless, tenants-in-common have been held to be liable as joint occupiers.[210]

2.117 Trespassers can also be in joint occupation. The authorities establishing the possibility of joint occupation have referred to or relied on the

---

208 Reg 3(9).
209 *NICV v Fermanagh Protestant Board of Education* [1969] 1 W.L.R. 1708 at p1728.
210 *R v Paynter* (1847) 7 QB 255, affd (1847) 10 QB 908, which concerned Putney bridge, the ownership of which was split between numerous tenants-in-common.

'similar and concurrent rights to the use of the property'[211] of the joint occupiers. However, even where the joint occupiers are trespassers and therefore have no legal right to be present, they can still be in joint occupation.[212]

2.118   The rules on joint occupation are particularly important when considering unincorporated associations. As the courts have stressed on numerous occasions, an unincorporated association cannot be in occupation of a hereditament because it is a nonentity.[213] It has no legal existence. Nor will the membership of the association automatically be in occupation.[214] It is therefore necessary to look for some individual or individuals who are in rateable occupation. Often this will be the trustees or committee of the association who may well have signed any lease or licence entitling the association to use the hereditament, and who have funds available to pay the bills.[215] It will however be a question of fact in each case whether any members of such a committee, and if so which members, are in occupation.

2.119   Joint occupation of the whole hereditament is different from a situation where two or more different people are in occupation of parts of the hereditament. In such cases it is long established that none of the parties is in occupation of the whole and therefore none of them is liable to pay rates on the whole.[216] This is true even if the various occupiers of parts of a hereditament are companies controlled by the same person or parent company.[217] Nor can they be liable for a part of those rates because the list will not contain any information as to the rateable value of the part occupied by each. This position is not altered by the provisions in the Non-Domestic Rating (Collection and Enforcement) (Miscellaneous Provisions)

211 *Griffiths v Gower RDC* (1972) 17 RRC 69 at p70, in which the court rejected an argument that joint occupation was not possible. See also *NICV v Fermanagh Protestant Board of Education* [1969] 1 W.L.R. 1708 at p1728, which makes reference to the 'joint right' of occupation.

212 *Westminster CC v Tomlin* [1989] 1 WLR 1287 at p1295.

213 *R v Brighton Justices, ex parte Howard* [1980] R.A. 222 at p225, *Verrall v Hackney London Borough Council* [1983] Q.B. 445 at p461–462, *Westminster CC v Tomlin* [1989] 1 WLR 1287 at p1292.

214 *Verrall v Hackney London Borough Council* [1983] Q.B. 445 at p462.

215 *Verrall v Hackney London Borough Council* [1983] Q.B. 445 at p461–462.

216 See *Allchurch v Hendon UAC* [1891] 2 QB 436.

217 *Tallington Lakes v Grantham Magistrates' Court* [2010] EWHC 3403 (Admin).

Regulations 1990.[218] These provisions were not intended to make any of those occupying a part of the hereditament liable for rates on the whole.[219] It follows that, unless and until the list is altered to show more than one hereditament, no one can be held liable for the rates as the 'occupier'.

Occupation of advertising rights and moorings

2.120 Specific statutory provision is made in respect of the occupation of advertisement hereditaments and multiple moorings.

2.121 There are two ways an advertisement can give rise to an entry in the list, and LGFA 1988 makes specific provision in respect of the occupation of both.

2.122 The first is as a freestanding 'advertising right', which is let out or reserved to someone other than the owner or occupier of land.[220] Advertising rights entered in the list as such are treated as occupied by the person 'for the time being entitled to the right'.[221]

2.123 The second is where land is used simply for advertisements or for a structure used to exhibit advertisements. In the second case, there is no need to rely on the specific statutory provisions on advertising rights to enter the advertisement structure in the list. Where such a hereditament is not occupied, the statute provides that it should be treated as occupied by 'the person permitting it to be so used or, if that person cannot be ascertained, its owner'.[222]

2.124 Although meters also attract special treatment in terms of list entries,[223] there is no corresponding provision in terms of occupation.

2.125 There is a power to enter multiple moorings in the same ownership as one hereditament.[224] Where this power has been exercised, then the owner, and no one else, is treated as being in occupation of those moorings.[225]

---

218 Discussed at para 2.114.
219 *Ford v Burnley Magistrates' Court* [1995] RA 205.
220 LGFA 1988, s64(2); discussed at 2.19.
221 LGFA 1988, s65(8).
222 LGFA 1988, s65(8A).
223 On which see para 2.19.
224 Non-Domestic Rating (Multiple Moorings) Regulations 1992, reg 2, and see para 2.14.
225 Non-Domestic Rating (Multiple Moorings) Regulations 1992, reg 3.

3

Liability to rates

3.1 The LGFA 1988 and its accompanying regulations determine the principle of liability to rates for both occupied and unoccupied hereditaments shown on the local list. The position of central list hereditaments is rather simpler and is dealt with separately below. The amount of liability in any case will depend on the rateable value of the hereditament and on the 'rate' or multiplier applied to that rateable value. The position as to liability set out in this chapter is of course subject to the various exemptions and reliefs discussed in the following chapter. Transitional relief is dealt with below because its application is effectively a question of calculation.

3.2 It should also be noted that death does not cancel a ratepayer's liabilities to rates. His executor or administrator becomes liable to pay sums which, if alive, he would have been due to pay.[1]

OCCUPIED LOCAL LIST HEREDITAMENTS

Principle of liability

3.3 Liability in respect of occupied hereditaments shown in a local list is governed very simply by s43(1) of LGFA 1988, which provides as follows:

(1) A person (the ratepayer) shall as regards a hereditament be subject to a non-domestic rate in respect of a chargeable financial year if the following conditions are fulfilled in respect of any day in the year—

---

1 Non-Domestic Rating (Collection and Enforcement) (Local List) Regulations 1989, reg 24.

    (a) on the day the ratepayer is in occupation of all or part of the hereditament, and

    (b) the hereditament is shown for the day in a local non-domestic rating list in force for the year.

3.4   Thus, where a hereditament shown in the list is occupied, the occupier will be liable to non-domestic rates.

3.5   The statute refers to a situation where the ratepayer occupies 'all or part of the hereditament'. If the ratepayer occupies all of the hereditament, then the situation is straightforward. It is possible for two or more people to be in occupation as joint occupiers; but not for two people to be in occupation of different parts of the hereditament.[2]

3.6   Where part of the hereditament is occupied, and the rest unoccupied, then the ratepayer is treated as being in occupation of the whole (and therefore liable to pay rates on the value of the whole). Ultimately the unfairness of this situation can be put right by altering the entry in the list so that it shows two or more smaller hereditaments. In the short term, however, there is provision for the amount of any liability to be reduced whilst the hereditament shown in the list is only partially occupied; this is discussed below with reference to the amount of the liability.[3]

Amount of liability

3.7   The amount of liability in a given year is calculated by aggregating the chargeable amounts for each chargeable day (that is, each day on which the conditions for liability are satisfied).[4]

3.8   The liability for an individual chargeable day is calculated according to the following formula:

$$\frac{A \times B}{C}$$

A = rateable value shown for the day in the local rating list;
B = non-domestic rating multiplier for the financial year;
C = number of days in the financial year.[5]

---

2 On joint and partial occupation, see paras 2.113–119.
3 See para 3.11 onwards.
4 LGFA 1988, s43(2)–(3).
5 The formula is set in LGFA 1988 s43(4) with definitions of the terms in s44.

3.9   The non-domestic rating multiplier is set by the Secretary of State in accordance with the provisions in LGFA 1988, schedule 7.[6] In the case of a 'special authority', it is set by the authority. The only current special authority is the City of London. At the time of writing, the Government is consulting on proposals to allow other authorities more autonomy over the multiplier. The multiplier once set can only be challenged by way of judicial review proceedings.[7]

3.10 In simple terms, therefore, the business rates payable by an occupier in a given year will (all other things being equal) be the product of the multiplier and the rateable value of the hereditament in question.

3.11 Where the hereditament is only part-occupied, the rateable value may be apportioned between the occupied and unoccupied parts under s44A of LGFA 1988. The billing authority can require that the valuation officer make and certify such an apportionment. A request can only be made when 'it appears to the authority that part of the hereditament is unoccupied but will remain so for a short time only'.[8]

3.12 The wording of this provision is prospective ('*will* remain so for a short time only . . .'). This may suggest that there is no power to make such a request retrospectively, which on the face of it is an odd conclusion. It is also not the view of the Government on the effect of this provision.[9] A further oddity of the provision on apportionment is that the apportionment has effect from the date when the hereditament became part-occupied (or from the date of the request where the new apportionment replaces an old apportionment). It would therefore seem that if a request is made in the final week of a long period of partial occupation, the resulting apportionment will have effect throughout the whole period.

3.13 It is not clear by what criteria a billing authority is to decide how long 'a short time' is in this context. The Welsh (but not the English) guidance suggests that billing authorities should adopt a uniform measure of what is a short time, 'taking account the prevailing commercial property market', and setting different periods for different types of

6 LGFA 1988, s56.
7 LGFA 1988, s138.
8 LGFA 1988, s44A(1).
9 See *Non-Domestic Rates: Guidance on Rate Relief for Charities and Other Non-Profit Making Organisations* (OPDM, 2002) at paragraph 8.2.7; and paragraph 8.2.8 in the 2004 Welsh guidance of the same title.

hereditament if 'local conditions merit it . . . e.g. if a particular sector of the local economy is weak'.[10]

3.14   The criteria for exercising the discretion to request an apportionment are also unclear. Government guidance makes the following suggestion:

> When exercising their discretion, authorities should have regard to the general rule that a person who occupied a part of a property is deemed to be in rateable occupation of the whole. Thus it is not intended that because part of a property is temporarily not used it should be taken out of rating. But, for example, where there are practical difficulties in occupying or vacating a property in one operation (perhaps because new accommodation to which the occupier is moving is not fully ready for occupation) and it is phased over a number of weeks or months, it would be reasonable to reduce the liability on that part of the property which is unoccupied. Similarly, where a building or buildings on a manufacturing site become temporarily redundant it might be reasonable to take the unoccupied part of the property into account rather than levy full rates on the whole property. Authorities should also bear in mind that such an approach may alleviate hardship in some circumstances.[11]

3.15   This appears to imply that an apportionment should only be made when part is unoccupied for reasons beyond the control of the ratepayer. Billing authorities may also have their own policies on when an apportionment will be requested.

3.16   The apportionment comes to an end where one of the following events has occurred:

(a)   the occupation of any of the unoccupied part of the hereditament;
(b)   the ending of the rate period[12] in which the authority requires the apportionment;

10   *Non-Domestic Rates: Guidance on Rate Relief for Charities and Other Non-Profit Making Organisations* (WAG, 2004) at paragraph 8.2.4.
11   *Non-Domestic Rates: Guidance on Rate Relief for Charities and Other Non-Profit Making Organisations* (OPDM, 2002) at paragraph 8.2.2; and compare the 2004 Welsh guidance of the same title at 8.2.2.
12   The term 'rate period' is not defined in LGFA 1988, but with reference to s3 of the General Rate Act 1967 would appear to relate to the period within which a rate (i.e. a multiplier in modern parlance) is set. As such, the 'rate period' will end on 31 March of each year.

(c) a further apportionment in relation to the same hereditament;

(d) the hereditament becoming completely unoccupied.[13]

3.17 The apportionment will only have an effect on the amount of liability if rates would not fall to be payable on the hereditament as a whole were it unoccupied. If no unoccupied rates would in general fall to be paid,[14] then the ratepayer simply pays occupied rates on that portion of the value which has been apportioned to the occupied part. This is achieved by using the apportioned value as 'A' in the formula above, in respect of all days on which the apportionment is effective.[15] In any other case, the apportionment will have no effect on the amount of any liability.[16]

3.18 In addition to the liability imposed under LGFA 1988, the Business Rate Supplements Act 2009 allows local authorities to levy a supplementary rate on hereditaments with a rateable value of over £50,000.[17] This Act has so far been implemented only in England. The supplement can only be levied to support a specific project.[18] Various procedural requirements must be met before the levy can be imposed, including a ballot with a majority both by number and rateable value approving the supplement.[19] The amount of the supplement cannot be more than 2p for every £1 of rateable value.[20]

UNOCCUPIED LOCAL LIST HEREDITAMENTS

3.19 Rates are payable by the owner of unoccupied hereditaments shown in the local list after three months of vacancy, or six months in the case of 'qualifying industrial hereditaments'.

---

13 LGFA 1988, s44A(5).

14 i.e. if the hereditament is either not a prescribed hereditament for these purposes under s45(1)(d) or would fall to be zero rated under s45A; see para 3.21 and Chapter 4 for more details on these provisions.

15 LGFA 1988, s44A(7).

16 LGFA 1988, s44A(8)–(9A). There is power under s45(4A) to prescribe that a different rate should be payable on the unoccupied portion, but that power has not been exercised.

17 Business Rate Supplements Act 2009, s12 and Business Rate Supplements (Rateable Value Condition) (England) Regulations 2009, reg 2.

18 Business Rate Supplements Act 2009, s3.

19 Business Rate Supplements Act 2009, s8.

20 Business Rate Supplements Act 2009, s14.

3.20   This result is achieved through a slightly convoluted statutory scheme.
        Section 45(1) of LGFA 1988 provides for liability where the following
        conditions are satisfied:

(a)   on the day none of the hereditament is occupied;
(b)   on the day the ratepayer is the owner of the whole of the
        hereditament;
(c)   the hereditament is shown for the day in a local non-domestic
        rating list in force for the year; and
(d)   on the day the hereditament falls within a class prescribed by the
        Secretary of State by regulations.

3.21   The determining factor is therefore whether the hereditament falls
        within a class prescribed by regulations. The relevant regulations in
        fact prescribe *all* 'relevant' hereditaments on the local list, subject
        to certain exceptions.[21] A 'relevant' hereditament for these purposes
        is defined as a hereditament comprised of any building or part of a
        building, together with land ordinarily used or intended for use for the
        purposes of the building or part.[22]

3.22   The main exceptions are those related to the time for which the her-
        editament has been unoccupied. A hereditament which has been
        unoccupied for a continuous period not exceeding three months, or six
        months in the case of a qualifying industrial hereditament, is exempt
        from unoccupied rates.[23] A 'qualifying industrial hereditament' is a
        hereditament (other than a retail hereditament) which is constructed
        or adapted for use in the course of a trade or business, and for purposes
        including manufacturing, processing goods, storage, mineral working
        and electricity generation.[24] This category has been held to extend to a

---

21   Non-Domestic Rating (Unoccupied Property) (England) Regulations 2008, reg 3;
     Non-Domestic Rating (Unoccupied Property) (Wales) Regulations 2008, reg 3.
22   Non-Domestic Rating (Unoccupied Property) (England) Regulations 2008,
     reg 2; Non-Domestic Rating (Unoccupied Property) (Wales) Regulations 2008,
     reg 2. This is not to be confused with the definition of a 'relevant' hereditament
     discussed at para 2.16.
23   Non-Domestic Rating (Unoccupied Property) (England) Regulations 2008, reg
     4(a)–(b); Non-Domestic Rating (Unoccupied Property) (Wales) Regulations
     2008, reg 4(a)–(b).
24   The full terms of the definition are in Non-Domestic Rating (Unoccupied
     Property) (England) Regulations 2008, reg 2; Non-Domestic Rating (Unoccupied
     Property) (Wales) Regulations 2008, reg 2.

surprisingly wide range of hereditaments; a sorting office has been held to come within it on the basis that mail was goods which when sorted was being subjected to a process and then stored.[25] On the other hand, 'storage' is not wide enough to include overnight parking of buses[26] or the housing of computers themselves used for data storage.[27] Although the requirement is that the hereditament be 'constructed or adapted' for the relevant purposes, the usual approach applied by the courts is to consider its actual previous use.

3.23   In calculating for how long a hereditament has been unoccupied, any period of occupation lasting less than six weeks is to be disregarded. The hereditament is treated as having been unoccupied throughout.[28] It is this provision which has given rise to rates-avoidance schemes involving the periodic occupation of otherwise empty hereditaments for six weeks and one day, so as to trigger the application of a further three- or six-month period of exemption.

3.24   The amount of liability in respect of an unoccupied hereditament is calculated in the same way as for an occupied hereditament (on which, see above).[29]

CENTRAL LIST HEREDITAMENTS

3.25   The central list includes the names of ratepayers. The determination of liability is therefore straightforward, as the person named in the list is liable to pay rates.[30] This is so whether the hereditament is occupied or unoccupied.

3.26   The amount of this liability is calculated in the same way as for occupied local list hereditaments (on which, see above).[31] The liability is payable to the Secretary of State rather than to a billing authority.

25 *Southwark LBC v Bellway Homes* [2000] RA 437.
26 *LB Barnet v London Transport Property* [1995] RA 235.
27 *Leda Properties Ltd v Kennet DC* [2003] RA 69.
28 Non-Domestic Rating (Unoccupied Property) (England) Regulations 2008, reg 5;
    Non-Domestic Rating (Unoccupied Property) (Wales) Regulations 2008, reg 5.
29 LGFA 1988, ss45, 46.
30 LGFA 1988, s54(1).
31 LGFA 1988, s54(2)–(7).

TRANSITIONAL RELIEF

3.27 Since 1990, schemes have been put in place to ameliorate the effect of a new list on rate liabilities. They have operated to moderate the suddenness of any change in liability consequent on the new list, whether the change was an increase or a decrease.

3.28 In respect of the period after 1 April 2010, there is a transitional scheme in place in relation to England, but not to Wales. It is set out in detailed technical regulations, which are not summarised here.[32] Those regulations apply to the period up to 31 March 2015, which in normal circumstance would have been the last day to which the 2010 list applied. The life of the 2010 list has been prolonged by two years, however. The regulations have not been amended to reflect this fact. Instead, the Government has allowed billing authorities to continue to provide transitional relief for hereditaments with rateable values up to £50,000, and indicated that it will reimburse them for the costs of doing so.[33]

32 Non-Domestic Rating (Chargeable Amounts) (England) Regulations 2009.
33 *Extension of Transitional Relief for small and medium properties – Guidance*, Jan 2015.

4

Exemptions and reliefs

4.1   There is a large number of exemptions and reliefs which may be available to ratepayers. There is little unifying principle in terms of which property or which ratepayers will be entitled to these.

4.2   If a hereditament is 'exempt' it is outside the scope of rating altogether. Most exemptions are provided for by schedule 5 of LGFA 1988, although there are still some common law exemptions (e.g. for highways and parks). Because exemption takes a property outside the scope of rating altogether, the way to establish that it exists is to seek an alteration of the list. Exemptions are dealt with first in this chapter.

4.3   'Relief' applies to reduce or extinguish the amount of rates that are payable in respect of a hereditament. It is applied by the billing authority. In some cases the entitlement to the relief is mandatory, in which case a billing authority's decision as to whether or not to apply it can be challenged when the billing authority attempts to recover the rates. Where the relief is available only at the discretion of the billing authority, however, then a decision to withhold relief can only be challenged by taking proceedings for judicial review.

4.4   The provisions specifically applicable to unoccupied property are dealt with in a distinct section, although unoccupied property may be entitled to another form of discount. In particular, it should be noted that the exemption provisions in schedule 5 of LGFA 1988 are to be applied to empty properties with reference to the position that 'it appears' will apply when the property is next in use or occupied.[1] This means that many unoccupied hereditaments will be exempt under these provisions.

---

1 LGFA 1988, sch 5 para 21.

AGRICULTURAL HEREDITAMENTS

4.5   Agricultural hereditaments have long been exempt from rating. The current legislation provides for the following to be exempt:

(1)   Agricultural land
(2)   Agricultural buildings
(3)   Fish farms.[2]

4.6   These are considered in turn below.

Agricultural land

4.7   Agricultural land is defined as follows in the LGFA 1988:[3]

(1)   Agricultural land is—

(a)   land used as arable, meadow or pasture ground only,
(b)   land used for a plantation or a wood or for the growth of saleable underwood,
(c)   land exceeding 0.10 hectares and used for the purposes of poultry farming,
(d)   anything which consists of a market garden, nursery ground, orchard or allotment (which here includes an allotment garden within the meaning of the Allotments Act 1922), or
(e)   land occupied with, and used solely in connection with the use of, a building which (or buildings each of which) is an agricultural building by virtue of paragraph 4, 5, 6 or 7 below.

(2)   But agricultural land does not include—

(a)   land occupied together with a house as a park,
(b)   gardens (other than market gardens),
(c)   pleasure grounds,
(d)   land used mainly or exclusively for purposes of sport or recreation, or
(e)   land used as a racecourse.

2 LGFA 1988, sch 5 paras 1 and 9.
3 LGFA 1988, sch 5 para 2.

4.8   Whether or not the definition applies will be a question of fact. For the most part its terms are self-explanatory. The following clarifications and illustrations do arise from the numerous decided cases, however.

4.9   'Land' does not for these purposes include buildings, because there are separate specific provisions on agricultural exemption for buildings.[4] In determining the use of land, future intended uses will not disqualify land which is currently used for agriculture.[5] However, where land is clearly in the process of being put into agricultural use, even if it is not clear what agricultural use in particular, it will still attract the exemption.[6]

4.10   In considering whether land is used 'only' as arable, meadow or pasture ground, it is possible to ignore minimal uses.[7] However, any other significant use will prevent this part of the definition from applying: the governing word is 'only'.[8] Similarly, when applying the exclusion in paragraph 2(2)(e) for land used as a racecourse, any use as a racecourse which is more than minimal will prevent the exemption from applying.[9]

4.11   The exclusion for 'land used mainly or exclusively for purposes of sport or recreation' has caused some confusion. It appears to imply that land used only partly (but not mainly or exclusively) for the purpose of sport or recreation would still be exempt. Does such land nevertheless fail to secure the exemption as it is not used 'only' as arable, meadow or pasture ground? The Court of Appeal has held that the word 'only' must not exclude land used for purposes of sport because this would 'render the latter part repugnant to what has gone before'.[10] After some vacillation,[11] the Lands Tribunal appears to have resolved the question by finding that in cases of sport or recreation, the use must fall within

---

4 *W & JB Eastwood Ltd v Herrod* (VO) [1971] AC 160 per Lord Reid at p167, on the predecessor provisions. Reaffirmed recently by the Court of Appeal in *Tunnel Tech Ltd v Reeves* (VO) [2015] EWCA Civ 718, at [64]–[66].

5 *Abercorn Estates Co v Edinburgh Assessor* (1935) SC 868, 6 DRA 7.

6 *National Pig Progeny Testing Board v Greenall* (VO) [1960] 1 WLR 1265.

7 *Watkins v Herefordshire Assessment Committee* (1935) 23 R&IT 304, 52 TLR 148.

8 *Meriden and Solihull RA v Tyacke* [1950] 1 All ER 939, (1950) 43 R &IT.

9 *Wimbourne and Cranborne RDC v East Dorset AC* [1940] 2 KB 420, confirmed by the House of Lords in *Hayes* (VO) *v Lloyd* [1985] 1 WLR 714.

10 *Garnett v Wand* (VO) (1960) 7 RRC 99.

11 *United Counties Agricultural Society v Knight* (VO) [1973] RA 13 and *Moore v Williamson* (VO) [1973] RA 172 both decide or imply that the exclusions have no effect on the interpretation of the exemption itself.

the exclusion if the exemption is to be lost. Use for sport or recreation that is more than minimal, but which does not meet the test in paragraph 2(2)(d), will accordingly not lead to the loss of agricultural exemption.[12]

4.12 The distinction between a 'market garden' and a 'nursery ground' within paragraph 2(1)(d) is significant when it comes to the treatment of associated buildings (on which, see below). Neither term is defined in LGFA 1988. The Court of Appeal has recently explained the difference between them in this way:

> what distinguishes a hereditament which is a market garden from one which is nursery ground is that there are produced on a hereditament which is a market garden such articles as fruit and vegetables[13] for sale and for consumption directly or indirectly by the public whereas the produce of a nursery is not suitable for, or not intended for, public consumption without some further process. Inevitably, there will some cases at the margins where the distinction between a market garden and nursery ground is not entirely straightforward. To that extent . . . the question is one of fact and degree.[14]

Agricultural buildings

4.13 LGFA 1988 contains a primary definition of an agricultural building.[15] This definition is also supplemented by further definitions relating to livestock buildings,[16] multiple and corporate occupiers,[17] and bee keeping,[18] all of which are considered in turn below.

---

12 *Eden (VO) v Grass Ski Promotions Ltd* [1981] RA 7. This decision is directly contrary to the same member's earlier decision in *Moore v Williamson (VO)* [1973] RA 172. However, it is to be preferred given that it is later in time and consistent with Court of Appeal authority in *Garnett v Wand (VO)* (1960) 7 RRC 99.
13 It should be noted that the growing of flowers has also been recognised as an ordinary part of 'market garden' use: see *Hood Bars v Howard* [1967] RA 50 per Willmer LJ at p60.
14 *Tunnel Tech Ltd v Reeves (VO)* [2015] EWCA Civ 718, at [49].
15 LGFA 1988, sch 5 para 3.
16 LGFA 1988, sch 5 para 5.
17 LGFA 1988, sch 5 paras 4 and 6.
18 LGFA 1988, sch 5 para 7.

4.14 Two points of application to all the different definitions are as follows. First, the term 'building'. This is to be construed broadly to reflect the nature of farm buildings, which may not always be of traditional construction. It cannot therefore be restricted to enclosures of brick or stonework, and timber poultry houses have been held to fall within it.[19] Furthermore, the statute provides that building in this context includes a separate part of a building.[20] Thus a part of a building can be exempt, providing that it is sufficiently defined to be 'a separate part'.

4.15 Second, several of the definitions require a building to be 'solely' used in one way or another. This contrasts with, for example, the provision on charitable relief which requires a hereditament to be 'wholly or mainly' used for charitable purposes. It seems therefore that the quantity of use is not relevant to the agricultural exemption. The only relevant matter is the purpose of the use that is actually being carried on.

4.16 Applying the normal meaning of the word 'solely', any use for a purpose other than that specified in the relevant definition will mean that the definition is not met. However, LGFA 1988 provides that in applying the test of 'sole use', 'no account shall be taken of any time during which [the building] is used in any other way, if that time does not amount to a substantial part of the time during which the building is used'.[21] The Court of Appeal has held that this provision only applies where the other use takes place in a 'definable period of time'.[22] If the other use is in effect mingled in with the qualifying use, the provision does not assist a ratepayer.

4.17 The existence of this statutory provision moderating the application of the 'sole use' test means that the maxim *de minimis non curat lex*[23] does not apply here.[24] As such, *any* other use, however minimal, which did not take place in a separate and definable period of time, will prevent the application of the exemption. This is a somewhat surprising conclusion and is arguably wrong. However, it is the view

---

19 *Shaw v Borrett (VO)* [1967] RA 90 at p94.

20 LGFA 1988, sch 5 para 8(4).

21 LGFA 1988, sch 5 para 8(3).

22 *Farmer (VO) v Buxted Poultry Ltd* [1991] RA 267 at p279–280. This part of the decision was not appealed to the House of Lords: [1993] AC 369 at p375.

23 The law is not concerned with trivial matters.

24 *Farmer (VO) v Buxted Poultry Ltd* [1991] RA 267 at p280. This part of the decision was not appealed to the House of Lords: [1993] AC 369 at p375.

of the Court of Appeal so must be applied until reversed by that or a higher court.

4.18   The primary definition of an agricultural building is as follows:[25]

> A building is an agricultural building if it is not a dwelling and—
>
> (a) it is occupied together with agricultural land and is used solely in connection with agricultural operations on that or other agricultural land, or
> (b) it is or forms part of a market garden and is used solely in connection with agricultural operations at the market garden.

4.19   The main part of this definition is applicable to buildings occupied and used with agricultural land. It therefore contains two key elements. The first relates to occupation, the second to use. Both have been considered in decisions of the House of Lords.

4.20   The first element is that the building be 'occupied together with agricultural land'. This requires the identification of a 'single agricultural unit' comprising both the land and the building. In applying this test, it is necessary to show that the land and buildings are in common occupation and that the activities carried on in both are jointly controlled or managed at the material times. These are necessary but not sufficient conditions. Beyond this, the main criterion is one of distance. Where the land and buildings are contiguous or nearly contiguous, it is submitted that the test will be met. Where the land and buildings are not contiguous, the cases indicate that they may still form a single agricultural unit. It is not correct to ask whether the two premises constitute a 'farm' in the ordinary sense.[26]

4.21   This first element does not apply to buildings used with or as market gardens. There may be no accompanying land or alternatively the buildings may 'form part of' a market garden (in the sense of a business enterprise) even if they are not 'occupied together with' agricultural land making up the market garden.

4.22   The second element is that the building is 'used solely in connection with agricultural operations on that or other agricultural land'. A building can only be said to be used 'in connection with agricultural

25 LGFA 1988, sch 5 para 3.
26 *Farmer (VO) v Buxted Poultry Ltd* [1993] AC 369.

operations' if the *use* of the building is ancillary or incidental to the *use* of the land, rather than the other way round.[27] Where the use of the land is ancillary to the use of the buildings, this part of the definition will not be satisfied. Finally, the use must be in connection with 'agricultural operations'. This phrase obviously includes operations by way of cultivating the soil or rearing livestock.[28] It has also been held to extend to 'operations reasonably necessary to make the product marketable or disposable to profit', but not so far as a separate business occupied with marketing or disposing of the product (such as a butcher's shop on a farm).[29] It will be a question of fact in each case whether a particular use falls within the definition or outside it.

4.23   The primary definition of an agricultural building has been supplemented by various more specialised provisions. Livestock and associated buildings are caught by the following:

> (1) A building is an agricultural building if—
>
> > (a) it is used for the keeping or breeding of livestock, or
> > (b) it is not a dwelling, it is occupied together with a building or buildings falling within paragraph (a) above, and it is used in connection with the operations carried on in that building or those buildings.[30]

4.24   In order for this provision to apply, the building must be 'surrounded by or contiguous to an area of agricultural land which amounts to not less than 2 hectares'.[31] LGFA 1988 makes provision for the calculation of this area and what must be excluded (essentially roads, railways, watercourses and other buildings).[32] Strangely, this land does not need to be in the same occupation as the building.[33]

---

27 *Eastwood v Herrod (VO)* [1971] AC 160 per Lord Reid at p168G, p169H–170A, Lord Morris at p174G, Lord Guest at p179F and per Viscount Dilhorne said at p181B. The actual result of this case was reversed by amendment of the legislation, but it remains good authority for the interpretation of the words in the current definition – see *Whitsbury Farm and Stud Ltd v Hemens (VO)* [1988] AC 601 at p612.

28 *Gilmore (VO) v Baker Carr* (1962) 9 RRC 240 at p243.

29 *Eastwood v Herrod (VO)* [1971] AC 160 at p169.

30 LGFA 1988, sch 5 para 5(1).

31 LGFA 1988, sch 5 para 5(4).

32 LGFA 1988, sch 5 para 5(5).

33 As the Court of Appeal observed in *Prior (VO) v Sovereign Chicken* [1984] 1 WLR 921 at p925.

4.25 Livestock is defined as including 'any mammal or bird kept for the production of food or wool or for the purpose of its use in the farming of land'.[34] In this context, the word 'include' has been held to cut down the normal meaning of the word 'livestock'.[35] As such, the term livestock in this context only covers animals kept for the purposes listed. It does not cover horses on a stud farm or pets. Buildings used for such animals will accordingly not come within this part of the exemption.

4.26 In order to benefit from this part of the exemption, a building must be used solely for the purposes mentioned, according to the approach to 'sole' uses discussed above. Alternatively, it can be used partly for that purpose and partly in the way envisaged by the primary part of the definition of agricultural buildings; that is to say, in connection with agricultural operations on accompanying agricultural land.[36]

4.27 The primary definition of an agricultural building has also been supplemented to take account of the position of farming syndicates and cooperatives. A building will be exempt if it is used solely in connection with agricultural operations carried out on agricultural land and is occupied jointly by the occupiers of the land, up to a maximum number of 25. Alternatively, they may occupy through intermediaries who are also occupiers of the relevant agricultural land.[37] Interestingly, in this case the building does not have to be 'occupied together with' the agricultural land in question. This means that it can be any distance from the relevant land as long as it is in the same occupation. This means that the provisions are potentially more generous than the primary definition. It seems however that this provision is only applicable to multiple occupiers and single occupiers cannot take advantage of it; the plural does not include the singular.[38]

4.28 The situation of a corporate occupier of a building is also dealt with in the legislation. A building used in connection with agricultural operations carried on on agricultural land will still attract exemption if it is occupied by a body corporate controlled by members (i.e. shareholders) who are occupiers of the relevant land.[39] In contrast to

---

34 LGFA 1988, sch 5 para 8(5).

35 *Whitsbury Farm and Stud Ltd v Hemens (VO)* [1988] AC 601 at p613–614.

36 LGFA 1988, sch 5 paras 5(2)–(3).

37 LGFA 1988, sch 5 para 4.

38 See *Prior (VO) v Sovereign Chicken* [1984] 1 WLR 921 decided on an earlier version of the statutory wording.

39 LGFA 1988, sch 5 para 7(1).

the provisions above, it appears here that the plural does include the singular such that this provision can apply where each member (shareholder) occupies one parcel of agricultural land. The agricultural land does not need to be in joint occupation for the exemption to apply.[40]

4.29   Like provisions apply to buildings which are used for purposes ancillary to livestock buildings.[41] These cover occupation jointly by an unincorporated group of individuals, occupation by representative members of a group and also occupation by a corporate body controlled by members (shareholders) who are the occupiers of the primary livestock buildings. In this case, it is sufficient if the building is occupied by one member of the group, despite the use of the word 'occupiers' in the legislation.[42]

Fish farms

4.30   LGFA 1988 contains specific provision for land and buildings used solely for or in connection with fish farming.[43] This provision is required because fish have been held not to be 'livestock'.[44]

PLACES OF WORSHIP

4.31   Three different types of hereditament are entitled to exemption under this heading:

     (1) Places of public religious worship
     (2) Church halls and similar buildings
     (3) Office and administration buildings of religious bodies.

4.32   These are dealt with in turn below. A general point to note is that the hereditament will be exempt 'to the extent that' it falls within one of these three categories. Partial exemption is therefore contemplated and it is not necessary in order to secure the exemption that the exempt building or part should form a separate hereditament. It does

---

40   *Farmer (VO) v Hambleton DC* [1999] RA 61.
41   LGFA 1988, sch 5 paras 7(2)–(5).
42   *Farmer (VO) v Hambleton DC and Buxted Chicken Ltd* [1999] RA 61.
43   LGFA 1988, sch 5 para 9.
44   *Cresswell (VO) v British Oxygen Co* [1980] 1 WLR 1556.

appear to be necessary, if there is to be partial exemption, that there be a 'definable part' of the hereditament that is so used.[45]

4.33 For the same reasons, it does not seem possible or appropriate to apply a partial exemption where a hereditament is used as a place of public religious worship etc. for only part of the *time*. Either a building (or part of a building) 'consists of' a place of public religious worship or church hall or it does not. It is also not clear how value would fall to be apportioned to different periods of use, which may overlap or coexist.[46]

Places of public religious worship

4.34 The full text of the exemption is as follows:

> (1) A hereditament is exempt to the extent that it consists of any of the following—
>
> (a) a place of public religious worship which belongs to the Church of England or the Church in Wales (within the meaning of the Welsh Church Act 1914) or is for the time being certified as required by law as a place of religious worship.

4.35 There are therefore three conditions to be met. First, the hereditament must be a place of religious worship. Second, it must be 'public'. Third, there is a formal requirement that the hereditament either belong to the established church or otherwise be certified as a place of religious worship under the Places of Worship Registration Act 1855.

---

45 *Gallagher (VO) v Church of JCLDS* [2008] UKHL 56 at [39] per Lord Hope, with whom a majority agreed. Note however the different opinion of Lord Mance at [55], who did not think there was any necessity for a separate definable part. The decision may not be strictly determinative of the point but is very likely to be followed given its source. It is consistent with the provisions on valuation of partially exempt hereditaments in s42(1)(c) and sch 6 para 2(1B) of LGFA 1988, which assume that a 'partially exempt hereditament' is a hereditament, some part of which is exempt.

46 This is in contrast the situation under the previous legislation, which gave clear directions for valuing hereditaments in these circumstances: General Rate Act 1967, s39(3).

4.36 In practice, the first requirement adds little to the third because only a 'place of meeting for religious worship' can be certified.[47] 'Religious worship' in this context is given a broad definition, in accordance with contemporary understanding of religion. 'Religion' is not to be confined to religions which recognise a supreme deity, and worship simply means religious services. Thus a church within the Church of Scientology met these criteria.[48]

4.37 In order to meet the second requirement, it is the place not the worship which must be public. A private place of worship (such as a family chapel) will not meet the test. Nor will a place of worship from which the public are generally excluded,[49] whether or not it can be described as 'private'.

Church halls and similar buildings

4.38 This exemption is as follows:

> (1) A hereditament is exempt to the extent that it consists of any of the following—
>
> > (b) a church hall, chapel hall or similar building used in connection with a place falling within paragraph (a) above for the purposes of the organisation responsible for the conduct of public religious worship in that place.

4.39 Paragraph (a) is a reference to the exemption for places of public religious worship. This ground of exemption therefore depends on the existence of an associated exempt place of religious worship.

4.40 There are three elements of this definition, each of which must be taken to narrow the field of exemption: first, a church or chapel hall or similar building; second, used in connection with a place of public religious worship; third, used for the purposes of the religious organisation.[50]

---

47 Places of Worship Registration Act 1855, s2.
48 *Regina (Hodkin and another) v Registrar General of Births, Deaths and Marriages* [2013] UKSC 77 at [34], [51], [57], reversing the earlier decision of the Court of Appeal on Scientology in *R v Registrar General ex p Segerdal* [1970] 2 QB 697.
49 *Gallagher (VO) v Church of JCLDS* [2008] UKHL 56 at [8], confirming the previous decision of the House of Lords in *Church of JCLDS v Henning (VO)* [1964] AC 420 and withholding the exemption from a Mormon temple only open to Mormons in 'good standing'.
50 *Swansea CC v Edwards (VO)* [1977] RA 209.

4.41 The words 'similar building' have been given a wide construction by the courts, such that they do not impose much of a practical obstacle to obtaining exemption. It is not an architectural test but about the character of the use.[51]

4.42 The words 'used in connection with' carry the same meaning here as they do in the exemptions for agricultural buildings.[52] That is to say, the church hall or similar building must be used for purposes ancillary to the place of public religious worship.[53]

4.43 Finally, it is notable that there is no requirement that the building should be wholly, mainly or solely used in that way, or for the purposes of the religious organisation. It has sometimes been considered that any significant use will suffice;[54] the better view is probably that the qualifying use has to be the primary use of the hereditament (or part of the hereditament) in order to attract the exemption.[55]

4.44 A churchyard will not be exempt under this provision as it is not a 'building'.

Office and administration buildings

4.45 The final form of religious exemption is as follows:

(2) A hereditament is exempt to the extent that it is occupied by an organisation responsible for the conduct of public religious worship in a place falling within sub-paragraph (1)(a) above and—

(a) is used for carrying out administrative or other activities relating to the organisation of the conduct of public religious worship in such a place; or

(b) is used as an office or for office purposes, or for purposes ancillary to its use as an office or for office purposes.

---

51 *West London Methodist Mission Trustees v Holborn BC* (1958) 3 RRC 86 at p91, in which the Divisional Court held that a Methodist centre arranged over seven floors and providing amongst other things a youth club, church socials, lunch club, crèche and playground was a 'similar building'.

52 See para 4.22.

53 *Gallagher (VO) v Church of JCLDS* [2008] UKHL 56 at [34]–[35].

54 See, e.g., *Westminster RC Diocese Trustee v Hampsher (VO)* [1975] RA 1 at p24.

55 *Gallagher (VO) v Church of JCLDS* [2008] UKHL 56 at [39] on the similar provisions of para 11(2).

4.46   Establishing that this ground of exemption applies again relies on the existence of a place of public religious worship benefiting from the relevant exemption. It extends to hereditaments used for organising public religious worship in such a place, but also to any office (whether or not it is associated with one or more places of religious worship or the activities which are conducted there). 'Office purposes' are defined in the Act.[56]

4.47   Once again, there is no requirement that the use in question should be the sole or main use of the hereditament. This has been interpreted as meaning that the use in question should be used 'primarily if not exclusively' for the purposes specified.[57]

HEREDITAMENTS DEDICATED TO THE PUBLIC

4.48   These hereditaments fall into three broad categories: public highways, parks and others.

Public highways

4.49   A public highway is not rateable. This is because it is dedicated to the public to such an extent that any beneficial use by the owner is impossible.[58] This doctrine was established, with reference to Putney Bridge, in *Hare v Putney Overseers*.[59]

Public parks

4.50   The *Putney* case was applied to a public park in *Lambeth Overseers v LCC*.[60] Exemption for parks is now covered by specific statutory provision in LGFA 1988.[61] This provision will normally be the starting point for establishing exemption. It states:

> (1) A hereditament is exempt to the extent that it consists of a park which—

---

56 LGFA 1988, sch 5 para 11(3).
57 *Gallagher (VO) v Church of JCLDS* [2008] UKHL 56 at [39].
58 *Kingston-upon-Hull v Clayton (VO)* [1963] AC 28 per Lord Tucker at p45, Lord Guest at p46.
59 (1881) 7 QBD 223.
60 [1897] AC 625.
61 LGFA 1988, sch 5 para 15.

(a) has been provided by, or is under the management of, a relevant authority or two or more relevant authorities acting in combination, and

(b) is available for free and unrestricted use by members of the public.

(2) The reference to a park includes a reference to a recreation or pleasure ground, a public walk, an open space within the meaning of the Open Spaces Act 1906, and a playing field provided under the Physical Training and Recreation Act 1937. . . .

(4) In construing sub-paragraph (1)(b) above any temporary closure (at night or otherwise) shall be ignored.

4.51 The term 'park' is partially defined in the provisions as including various sorts of land falling within two of the principal statutes[62] under which local authorities hold land used as a park. The definition of an 'open space' within the Open Spaces Act 1906 refers to various types of open land of which no more than one-twentieth is covered with buildings.[63] From this it can be inferred that the presence of some buildings does not prevent land from becoming a park; equally, however, an 'essential characteristic' of a park is that it is in the open air.[64]

4.52 As with other exemptions,[65] the words 'to the extent that' imply that a hereditament may be partially exempt. This is rare in practice. In the *Lambeth Overseers*[66] case itself, which first established the exemption for public parks, a mansion house and refreshment rooms in Brockwell Park were held to be part of the park and therefore exempt. A similar approach has prevailed in the application of the exemption both before[67]

---

62 The other is s175 of the Public Health Act 1875. Many parks and recreation grounds are also held under specific local Acts.

63 Open Spaces Act 1906, s20.

64 *Smith (VO) v St Albans City and DC* [1978] RA 147 at p157, holding that an indoor heated swimming pool could not be a 'park' for this reason.

65 See on exemptions for religious buildings at paras 4.32–33.

66 [1897] AC 625.

67 *Sheffield Corpn v Tranter (VO)* [1957] 1 WLR 843, in which the Court of Appeal held that a refreshment pavilion was held to be 'an essential amenity of a public park' and therefore to be exempt even though let out to a caterer: p854.

and after[68] the statutory provision was enacted. It is a question of fact whether the apparently separate 'part' is in fact still an inherent and essential part of the park, providing an amenity for it.[69]

4.53 The requirement for free and unrestricted use by members of the public does not preclude some regulation and charging for entry. It is again a question of fact whether that regulation and charging has passed the point at which use can no longer be said to be free and unrestricted. In practice, the amount of income obtained by the public authority may well be treated as determining on which side of the line particular parks fall.[70]

4.54 In order to benefit from the statutory exemption, a park must be provided by or under the management of a relevant authority as defined in LGFA 1988.[71] If a hereditament is not owned by such an authority, it cannot benefit from the exemption however much it may resemble a public park. In those circumstances, it may be necessary to have recourse to the pre-existing doctrine of exemption as set out in the *Lambeth Overseers* case and later cases.[72] In doing so it will be necessary to address the difficult question of whether the land has been perpetually 'dedicated' to public use.

Other hereditaments

4.55 There have been various attempts to extend the principle in *Lambeth Overseers* to other sorts of hereditament, such as museums[73] and libraries.[74] These have not, on the whole, been successful. The doctrine

---

68 *Oxford CC v Broadway (VO)* [1999] RA 169, in which an open-air swimming pool was held to be exempt as part of a public park.

69 *Sheffield Corpn v Tranter (VO)* [1957] 1 WLR 843, at p855.

70 In *North Riding of Yorkshire CVC v Redcar Corpn* [1943] 1 KB 114 the council made a significant profit from facilities on the foreshore, which were held rateable. In *Burnell (VO) v Downham Market* [1952] 2 QB 55 the rent paid by a federation of cricket and football clubs for the privilege of running the field and charging for admission for about 40 hours a year did not negate 'free and unrestricted' access; this facility was run at a loss by the council.

71 LGFA 1988, sch 5 para 15(3).

72 This is what the Lands Tribunal did in *Fenwick (VO) v Peak Park Joint Planning Board* [1983] RA 131, although concluding in the end that neither basis of exemption was satisfied.

73 *Sir John Soane's Museum Trustees v St Giles-in-the-Fields and St George's Bloomsbury Joint Vestry* (1900) 83 LT 248 (in which the submissions of a certain Mr Ryde were strongly rejected by the court), *Bowes Museum and Park Trustees v Cutts (VO)* (1950) 43 R&IT 881 and 900, and more recently *Kingston-upon-Hull Corpn v Clayton (VO)* [1963] AC 28.

74 *Liverpool Corpn v West Derby Union* (1905) 92 LT 467.

is therefore probably best seen as a particular exception from rating for public parks, with limited application to other forms of publicly owned and/or operated hereditaments.

HEREDITAMENTS USED FOR DISABLED PEOPLE

4.56 LGFA 1988 provides for exemption of various different types of hereditament used for or by disabled people. These provisions have been criticised as being 'hardly intelligible' and of failing to achieve a fair or rational provision of exemption for property used by disabled people.[75] The exemption is provided as follows:[76]

> (1) A hereditament is exempt to the extent that it consists of property used wholly for any of the following purposes—
>
>> (a) the provision of facilities for training, or keeping suitably occupied, persons who are disabled or who are or have been suffering from illness;
>> (b) the provision of welfare services for disabled persons;
>> (c) the provision of facilities under section 15 of the Disabled Persons (Employment) Act 1944;
>> (d) the provision of a workshop or of other facilities under section 3(1) of the Disabled Persons (Employment) Act 1958.

4.57 As with other exemptions, the words 'to the extent that' suggest that partial exemption is possible.[77] The inclusion of the word 'wholly' makes clear that the partial exemption can only apply to a defined part of the hereditament which is used exclusively for the purposes listed.

4.58 The definition of a person with a disability follows that set out in the Equality Act 2010: i.e. a physical or mental impairment which has a substantial and long-term effect on that person's ability to carry out normal day-to-day activities.[78] Illness is defined with

---

75 *Evans (VO) v Suffolk CC* [1997] RA 120 at p135.
76 LGFA 1988, sch 5 para 16.
77 See discussion in relation to religious hereditaments at 4.32–33.
78 LGFA 1988, sch 5 paras 16(1A)–(2). In Wales there is power to make regulations providing that persons in specified categories are or are not to be treated as disabled: Social Services and Well-being (Wales) Act 2014. No such regulations have yet been made.

reference to the National Health Service Act 2006; i.e. as including 'any disorder or disability of the mind and any injury or disability requiring medical or dental treatment or nursing'. The definition is not broad enough to cover all those who are suffering mental distress for whatever reason.[79]

4.59   The somewhat archaic expression 'facilities for training, or keeping suitably occupied' is to be construed together so as to give the sense of 'training or occupation'. As such holiday or recreational facilities do not fall into the category of facilities for keeping disabled people 'suitably occupied'.[80] Nor do facilities provided for a profitable trade or business carried on by disabled people for the purpose of profit.[81]

4.60   The 'provision of welfare services for disabled persons' is given a particular definition. In England it relates to services or facilities of the sort that could have been provided by a local authority under s29 of the National Assistance Act 1948.[82] In Wales it relates to services or facilities which a local authority can provide under ss35 and 36 of the Social Services and Well-being (Wales) Act 2014.[83] The English definition has been held to exclude residential facilities[84] and drug-counselling facilities.[85] The definition applicable in Wales is likely to be found to exclude the latter, but not the former.[86]

---

79 *Samaritans of Tyneside v Newcastle upon Tyne CC* [1985] RA 219.

80 *Chilcott (VO) v Day* [1995] RA 285 at p290.

81 *O'Kelly v Davey (VO)* [1996] RA 238 at p243–244.

82 This section allowed a local authority to 'make arrangements for promoting the welfare' of disabled persons. However, it did not extend to anything which fell to be provided by the NHS; i.e. services for the diagnosis and treatment of illness, including medical, dental, nursing and ambulance services.

83 LGFA 1988, sch 5 para 16(4).

84 *Jewish Blind Trustees v Henning (VO)* [1961] 1WLR 24 at p33; approved on this point by the House of Lords in *Vandyk v Oliver (VO)* [1976] AC 659. There was specific provision in s29 for hostels provided in connection with workshops, but otherwise no power to provide residential accommodation.

85 *Reid (VO) v Barking, Having and Brentwood Community Health Care Trust* [1997] RA 385; these were facilities for the diagnosis and treatment of illness.

86 The Social Services and Well-being (Wales) Act 2014 does not permit the provision of services required to be provided under a health enactment, unless it is incidental or ancillary: s47. In terms of residential facilities, however, these would seem to be covered by the terms of the Act as long as they would not fall to be provided under the Housing (Wales) Act 2014.

4.61 The final two paragraphs of the exemption deal with premises provided for the employment of disabled people. Where a hereditament or part is wholly used for facilities provided under the statutory powers mentioned, it will be exempt.

SOVEREIGN AND DIPLOMATIC IMMUNITY

4.62 The exemption from rating of Crown property was abolished by the Local Government and Rating Act 1997.[87] Crown property is no longer exempt from rating.

4.63 Foreign sovereigns and sovereign states remain immune from proceedings in the domestic courts. As such there is no way of enforcing liability to rates against them.[88] Hereditaments occupied by visiting forces and as headquarters also benefit from a specific statutory exemption.[89]

4.64 The position in respect of diplomats is governed by the Diplomatic Privileges Act 1964, which gives effect in domestic law to certain articles of the Vienna Convention on Diplomatic Relations.

4.65 In general, embassies and the ambassador's residence will be exempt from rates as 'the premises of a diplomatic mission'. The premises of a diplomatic mission are defined as:

> the buildings or parts of buildings and the land ancillary thereto, irrespective of ownership, used for the purposes of the mission including the residence of the head of the mission.[90]

4.66 The sending State and head of the mission are exempt from 'all national, regional or municipal dues and taxes' in respect of such premises.[91]

4.67 A diplomatic agent is also exempt from all dues and taxes on 'private immovable property' held on behalf of the sending State for the purposes of the diplomatic mission.[92] There is a similar immunity from

---

87 See LGFA 1988, s65A.
88 The immunity of sovereigns and heads of State in this regard is preserved specifically by the State Immunity Act 1978, s20(4).
89 LGFA 1988, sch 5 para 19A.
90 Diplomatic Privileges Act 1961, sch 1 art 1.
91 Diplomatic Privileges Act 1961, sch 1 art 23.
92 Diplomatic Privileges Act 1961, sch 1 art 34(b).

civil proceedings in respect of the same.[93] These same exemptions apply to non-national members of the diplomat's family forming part of his household[94] and to administrative and technical staff and their families, as long as they are not nationals nor permanently resident in the country, and are acting in the course of their duties.[95]

4.68   It follows that diplomats or others associated with a diplomatic mission do not enjoy immunity from liability to non-domestic rates in respect of independent business ventures which are not connected with the mission.

4.69   There are similar exemptions for consular premises[96] and to the premises of international organisations.[97] Diplomatic and consular immunity from taxation can be waived[98] or in certain circumstances withdrawn.[99]

MISCELLANEOUS EXEMPTIONS

Property of Trinity House

4.70   Trinity House is a charity dedicated to safeguarding shipping and seafarers, providing education, support and welfare to the seafaring community with a statutory duty as a General Lighthouse Authority to deliver a reliable, efficient and cost-effective aid-to-navigation service for the benefit and safety of all mariners. A hereditament is exempt to the extent that it belongs to or is occupied by Trinity House and consists of:

   (a)   a lighthouse;
   (b)   a buoy;
   (c)   a beacon;
   (d)   property within the same curtilage[100] as, and occupied for the purposes of, a lighthouse.[101]

---

93 Diplomatic Privileges Act 1961, sch 1 art 31(1)(a).
94 Diplomatic Privileges Act 1961, sch 1 art 37(1).
95 Diplomatic Privileges Act 1961, sch 1 art 37(2).
96 Consular Relations Act 1968, sch 1.
97 Conferred by orders made under the International Organisations Act 1968, s4.
98 Diplomatic Privileges Act 1961, sch 1 art 32.
99 Diplomatic Privileges Act 1961, s3; Consular Relations Act 1968, ss2 and 3.
100 For a discussion on the meaning of 'curtilage' see paras 2.29–30.
101 LGFA 1988, sch 5 para 12(1).

4.71   No other hereditament or part belonging to or occupied by Trinity House is exempt, despite a general exemption from 'taxes, duties or rates' provided by s221 of the Merchant Shipping Act 1995.[102]

Sewers

4.72   LGFA 1988 provides for exemption for sewers as follows:

> (1)   A hereditament is exempt to the extent that it consists of any of the following—
>
> (a)   a sewer;
> (b)   an accessory belonging to a sewer.
>
> (2)   'Sewer' has the meaning given by section 343 of the Public Health Act 1936.
> (3)   'Accessory' means a manhole, ventilating shaft, pumping station, pump or other accessory.

4.73   The Public Health Act 1936 in turn defines a 'sewer' and a 'drain' as follows:

> '*sewer*' does not include a drain as defined in this section but, save as aforesaid, includes all sewers and drains used for the drainage of buildings and yards appurtenant to buildings;
>   '*drain*' means a drain used for the drainage of one building or of any buildings or yards appurtenant to buildings within the same curtilage.

4.74   The effect of these provisions is that all sewers and drains will be exempt, except drains which are used for the drainage of one building or several buildings or yards within the same curtilage. The function of a sewer is to convey or transport sewage from the point of origin to the point of discharge.[103] A sewer continues until it reaches the point of disposal of the waste it is conveying.[104] 'Accessory' in the phrase 'other accessory' is to be given its ordinary meaning of something joined to

---

102 LGFA 1988, sch 5 para 12(2).
103 *Fife County Council v Fife Assessor* [1965] RA 373 per Lord Fraser at p377.
104 *Gudgion (VO) v Erith BC* (1961) 8 RRC 324 at p326.

another but subordinate, as auxiliary, or dependent upon it.[105] If items by their location and purpose are in fact part of sewage treatment works or are for the disposal of sewage, they will not be held to be accessory to the sewer.[106]

Drainage authorities

4.75   A hereditament is exempt to the extent that it consists of any of the following:

   (a)   land which is occupied by a drainage authority and which forms part of a main river or of a watercourse maintained by the authority;

   (b)   a structure maintained by a drainage authority for the purpose of controlling or regulating the flow of water in, into or out of a watercourse which forms part of a main river or is maintained by the authority;

   (c)   an appliance so maintained for that purpose.[107]

4.76   The Environment Agency is the 'drainage authority' for England; in Wales the relevant bodies are the Natural Resources Body for Wales or any internal drainage board. 'Main river' and 'watercourse' are defined in the drainage legislation.

Air raid protection works

4.77   Air raid protection works are exempt from rating, as long as they are not occupied for any other purpose.[108]

Swinging moorings

4.78   'Swinging' moorings are those comprising only a buoy secured by a weight on the sea or river bed, which are designed to be raised from the bed from time to time. They are exempt.[109]

---

105 *Jones (VO) v Eastern Valleys (Monmouthshire) JSB (No 2)* (1960) 6 RRC 387 at p390.
106 See, e.g., *Hoggett (VO) v Cheltenham Corpn* [1964] RA 1, *Fife County Council v Fife Assessor* [1965] RA 373.
107 LGFA 1988, sch 5 para 14(1).
108 LGFA 1988, sch 5 para 17.
109 LGFA 1988, sch 5 para 18.

River crossings

4.79 An occupied hereditament is exempt to the extent that it consists of a fixed road crossing over a watercourse.[110] 'Fixed road crossing' is defined expansively enough to be able to encompass an aerial ropeway along which a carriage runs.[111] There is a specific exclusion for floating bridges; i.e. ferries operating between fixed chains.[112] 'Appurtenances' of the crossing are also exempt.

4.80 It will be observed that, where the bridge carries a public highway, it is likely to be exempt in any event as a hereditament dedicated to the public.[113] This provision will therefore only be required if the river crossing does not for some reasons come within the scope of that exemption.

Property used for road charging schemes

4.81 Toll roads are exempt from rating, as are hereditaments used solely for or in connection with the operation of them (except office buildings).[114]

Enterprise zones

4.82 LGFA 1988 contains a total exemption for property situated in enterprise zones designated under the Local Government, Planning and Land Act 1980.[115] These were designated for a period of ten years, and the last one was designated on 21 October 1996,[116] so there are now no such enterprise zones in existence.

4.83 This statutory exemption is not to be confused with the more recent establishment of 'enterprise zones' in 2012 onwards. The Government has committed to fund discretionary relief granted under LGFA 1988 s47[117] in these zones up to certain amounts and in certain circumstances, which vary according to the zone in question.[118]

---

110 LGFA 1988, sch 5 para 18A.
111 *Griffin v Sansom (VO)* [1996] RA 454.
112 LGFA 1988, sch 5 para 18A(3).
113 See para 4.49.
114 LGFA 1988, sch 5 para 18B.
115 LGFA 1988, sch 5 para 19.
116 The Tyne Riverside Enterprise Zone, designated by SI 1996/2435.
117 For further discussion of s47 see 4.107 onwards.
118 See Business Rates Information Letters 5/2012, 6/2012.

RELIEF FOR CHARITIES AND LIKE ORGANISATIONS

Requirements for relief

4.84 Before 1955 there was no statutory provision for the granting of relief from rates to charities and relief was awarded on an informal basis. The Rating and Valuation Act 1955 first made statutory provision for charitable relief; subsequent legislation has narrowed the application of such relief. The reason for this restriction was to ensure that charitable relief was only awarded to charities making 'use of the building which is substantially and in real terms for the public benefit, so as to justify exemption from ordinary tax'.[119] This rationale should be borne in mind when applying the reliefs.

4.85 The occupier of a hereditament is entitled to 80 per cent relief (sometimes referred to as 'mandatory' relief) in respect of any day where two conditions are satisfied: (1) the ratepayer is a charity or trustees for a charity; (2) the hereditament is wholly or mainly used for charitable purposes (whether of that charity or of that and other charities).[120] A similar relief is applicable under the same provision where the ratepayer is a registered community amateur sports club. Where the provisions for mandatory relief are satisfied, the ratepayer can also apply to the billing authority for discretionary relief as to the remainder of the liability, and there are further powers to grant discretionary relief for institutions which are similar in character to charities.[121]

Identity of occupier

4.86 A charity is defined in LGFA 1988 as an institution or other organisation established for charitable purposes only or any persons administering a trust established for charitable purposes only.[122] It will generally be easy to ascertain whether an organisation is charitable by checking whether it is registered with the Charities Commission. If it is a registered charity then it is conclusively presumed to be established for charitable purposes only,[123] and so can be regarded as a charity for

---

119 *Public Safety Charitable Trust v Milton Keynes Council* [2013] EWHC 1237 (Admin) per Sales J at [34].
120 LGFA 1988, s43(6).
121 LGFA 1988, s47. See para 4.114 onwards.
122 LGFA 1988, s67(10).
123 Charities Act 2011, ss1(1) and 37(1).

the purposes of the LGFA 1988. If it is not so registered then it may still be charitable for the purposes of the LGFA 1988 providing it meets the definition.

4.87   The Charities Act 2011 defines what is meant by a 'charitable purpose'. In order to be charitable, a purpose must be: (a) included in the list set out in the Charities Act 2011; and (b) for the public benefit.[124] The list includes such familiar charitable purposes as 'the prevention or relief of poverty' and 'the advancement of religion', as well as more modern additions such as 'the advancement of environmental protection or improvement'. It also includes a catch-all category of purposes which are 'analogous to, or within the spirit of' any of the listed purposes.[125] The public benefit requirement, it is to be noted, is to be applied to the purposes of the charity, not its activities. However, the courts – wary of the 'skill of Chancery draftsmen' in devising purposes which apparently serve the public benefit – have indicated that it is sometimes permissible to look at the activities of a 'charity' to see if its purposes are indeed charitable.[126] In the vast majority of cases, however, there will be little room for dispute over whether an occupier is a charity or not.

4.88   Similarly, it will be easy to establish if the occupier is a registered community amateur sports club. The register is maintained by HMRC and can easily be checked online. Unlike charities, there is no provision for an *unregistered* community amateur sports club to benefit from rating relief.

Use of the hereditament

4.89   The requirement that the hereditament be 'wholly or mainly used' for charitable purposes (or for the purposes of a community amateur sports club) has caused more controversy. The courts have held that, in order for this requirement to be satisfied, both the purpose and the extent of the use must be considered.[127]

4.90   The starting point for assessing whether the purpose of the use satisfies the definition will be the charitable purposes of the occupying charity. The statutory wording means that at least some of the use

124 Charities Act 2011, s2(1).
125 Charities Act 2011, s3(1).
126 *Attorney-General v Ross* [1986] 1 WLR 252 at p263. See also *Southwood v Attorney-General* (2000) (unreported) per Chadwick LJ at para 5.
127 *Kenya Aid Programme v Sheffield City Council* [2013] EWHC 45 (Admin), *PSCT v Milton Keynes Council* [2013] EWHC 1237.

must be for the purposes of that charity specifically.[128] On the face of it, it might be thought that any use by a charity was use for the charitable purposes of that charity. It is clear from the terms of the section, however, that not all use by a charity will be regarded as use for charitable purposes. The House of Lords in *Oxfam v Birmingham CC*[129] observed that the line had to be drawn somewhere. In that case it was held that the line was to be drawn between user for purposes directly related to the achievement of the objects of the charity, on the one hand, and user for the purpose of getting in, raising or earning money for the charity, on the other.[130]

4.91 Use which is not directly related to the charitable purposes of the charity will therefore not be use for charitable purposes; the most common example is use for revenue raising. Any use for such purposes, or for any other non-charitable purposes, should be discounted. It should be noted that charity shops used wholly or mainly for the sale of donated goods benefit from a specific statutory provision which makes them charitable even though other revenue-raising activities are not.[131]

4.92 Sometimes it will not be easy to discern the purpose of a use. Where the charity is occupying through an agent or employee, it is the perspective of the charity that should be considered, not that of the agent or employee.[132] Similarly, a shop selling goods made by beneficiaries of the charity may be held to be used for a charitable purpose even though it also serves the purpose of revenue raising.[133] Where, however, a charity is occupying premises as part of a larger rates-avoidance scheme and generating revenue thereby, the High Court has suggested that it may be possible to conclude that this amounts to a separate and non-charitable purpose of the occupation.[134] In such cases the question

---

128 *Preston City Council v Oyston Angel Charity* [2012] EWHC 2005 (Admin); [2012] RA 357.
129 [1976] AC 126.
130 *Oxfam v Birmingham CC* [1976] AC 126 per Lord Cross at p146D–E.
131 LGFA 1988, s64(10). This provision was enacted to reverse the effect of the *Oxfam* decision.
132 *Glasgow Corporation v Johnstone* [1965] AC 609.
133 See, e.g., *Belfast Association for Employment of Industrious Blind v Valuation Commissioner for Northern Ireland* [1968] NI 21. Contrast with the decision in *RSPB v Brighton BC* [1982] RA 33, where the court decided that the revenue-raising purpose was the main purpose.
134 *Kenya Aid Programme v Sheffield City Council* [2013] EWHC 45 (Admin) at [39].

of what the purposes are and which purpose is the 'main' purpose will need to be assessed on the particular facts.

4.93   Assessing the quantity of use is also very much a question of looking at the particular facts. It must be remembered, however, that there is no requirement that the premises are 'necessary' for the charity, nor that they are used in a particularly efficient way; it is the actual use being made of the premises that must be considered.[135] The requirement is that the premises be 'wholly or mainly used' for charitable purposes. It is notable in this context that premises can be mainly *unused*,[136] and it is therefore not appropriate to take into account as a material consideration the fact that the only use going on is a charitable one.[137] Even if that is the only use going on, it must still be demonstrated that the premises are 'mainly' used.

4.94   In some cases it may be sufficient to establish that over 50 per cent of the premises are used for charitable purposes in order to satisfy the test. That simplistic approach does not take account of other factors which may be relevant, however, such as:

(1)   the nature of the premises: a different amount and pattern of use might be expected of an office compared to a sports stadium, for example;[138]

(2)   the duration of the use, bearing in mind that entitlement to the relief is assessed daily;[139]

(3)   where premises are used for multiple purposes, the relative importance to the charity of the different purposes.[140]

4.95   Furthermore, the courts have indicated that floor-space calculations designed to determine whether the premises are 'mainly' used for a particular purpose must be treated with caution. Such calculations will

---

135 *Kenya Aid Programme v Sheffield City Council* [2013] EWHC 45 (Admin) at [36]–[37].

136 *Public Safety Charitable Trust v Milton Keynes* [2013] EWHC 1237 (Admin) at [34].

137 *South Kesteven District Council v Digital Pipeline Limited* [2016] EWHC 101 (Admin) at [27].

138 *Public Safety Charitable Trust v Milton Keynes* [2013] EWHC 1237 (Admin) at [37], *English Speaking Union v Edinburgh City Council* [2009] CSOH 139 at [12]–[13].

139 On this, see discussion of LGFA 1988, s67 at 2.57.

140 See, e.g., *Wynn v Skegness UDC* [1967] 1 WLR 52 at p62F.

simply be one factor in the overall decision, which remains a 'broad brush' exercise. They will not be determinative. Particular care must be taken where it is not possible to identify specific defined areas which are or are not in use.[141]

RELIEF FOR SMALL AND RURAL BUSINESSES

4.96   The provisions in respect of England and Wales have diverged significantly in respect of small and rural businesses.

4.97   In England, there remain separate provisions for small businesses[142] and rural businesses.[143] In Wales there is one set of provisions granting relief to small businesses, which does not differentiate between rural and other areas.[144] All of these reliefs take effect, like mandatory charitable relief,[145] by modifying the formula under which liability to rates is calculated. As such there are rules as to which takes priority when several are engaged. These are set out below with regard to England and Wales separately.

4.98   In England, small business relief applies if the only hereditament occupied by a ratepayer falls below a certain rateable value.[146] If the ratepayer occupies multiple hereditaments, the relief is not available.[147] Where this relief applies, there are two effects. The first is that the 'small business multiplier' is used in place of the normal multiplier. The small business multiplier is set by the Government each year along with the normal multiplier and is normally around 1p lower. This has the effect of reducing the liability for small businesses by a corresponding amount per pound of rateable value. The second effect

---

141   *South Kesteven District Council v Digital Pipeline Limited* [2016] EWHC 101 (Admin) at [25]–[26].

142   LGFA 1988, s43(4A) and the Non-Domestic Rating (Small Business Rate Relief) (England) Order 2012.

143   LGFA 1988, s43(6A).

144   LGFA 1988, s43(4A) and the Non-Domestic Rating (Small Business Relief) (Wales) Order 2015.

145   See para 4.85.

146   Non-Domestic Rating (Small Business Rate Relief) (England) Order 2012, art 2: the current levels are £25,499 in Greater London or £17,999 outside Greater London.

147   Non-Domestic Rating (Small Business Rate Relief) (England) Order 2012, art 3(4) and 4(5).

is that hereditaments of particularly low value attract further relief. For hereditaments under £6,000 in value, the relief is 50 per cent. For hereditaments over £12,000, there is no further relief. In between those values there is a sliding scale.[148] These provisions have been temporarily modified such that until 31 March 2017 hereditaments below £6,000 in value effectively receive 100 per cent relief; the sliding scale is affected accordingly.[149] This temporary measure may well be extended again.

4.99 In England there are also separate provisions for certain hereditaments in rural settlements. Rural settlements are those which appear on a rural settlement list maintained by the billing authority. This list is required to show the boundaries of all settlements with a population of less than 3,000, which fall within areas designated as 'rural' by the Secretary of State.[150] Hereditaments falling within these settlements can benefit from a 50 per cent relief[151] if they meet two conditions. The first is that the hereditament has a value falling below certain prescribed levels.[152] The second is that it falls within one of the categories provided for, of which there are five: qualifying general store;[153] qualifying food store;[154] qualifying post office;[155] public house; and petrol filling station.[156] These terms are defined in detail in the legislation but in essence relate to hereditaments which are used wholly or mainly for the purpose specified. In respect of a general store or post office (but not a food store), the hereditament to qualify must be the only hereditament of that sort in the rural settlement.

4.100 There are thus two categories of relief applicable in England: for small businesses and for rural settlements. Where both apply to a hereditament, the rural settlement relief of 50 per cent takes priority over the

---

148 Non-Domestic Rating (Small Business Rate Relief) (England) Order 2012, art 3.
149 Non-Domestic Rating (Small Business Rate Relief) (England) Order 2012, art 4.
150 LGFA 1988, s42A–42B.
151 This is the effect of the formula in LGFA 1988 s43(6A).
152 Non-Domestic Rating (Rural Settlements) (England) Order 1997, art 3, as amended from time to time. The current limits are £12,500 for public houses and petrol filling stations and £8,500 for any other hereditament.
153 LGFA 1988, s43(6C).
154 LGFA 1988, s43(6CA).
155 LGFA 1988, s43(6D).
156 Non-Domestic Rating (Public Houses and Petrol Filling Stations) (England) Order 2001, art 2.

small business relief.[157] Where mandatory charitable relief of 80 per cent applies, that takes priority over the others.[158]

4.101   In Wales, the system is considerably simpler. The elaborate mechanism of rural settlements and small business multipliers has been disposed with. Instead there is one set of hereditaments which will benefit from relief, whether they are in rural or other areas.[159] This is also described as small business relief, but it is important to observe that the set of hereditaments which benefit from it is different from that in England.

4.102   In order to benefit from small business relief in Wales, a hereditament must have a rateable value of less than £12,000. It must also satisfy one of four sets of conditions, which are described as follows, and in each case set out in more detail in the relevant order:

(1) rateable value conditions (for hereditaments with rateable value below £7,800);[160]

(2) child-care conditions (for hereditaments wholly used for providing registered child care);[161]

(3) post office conditions (for post offices);[162]

(4) retail premises conditions (for hereditaments wholly or mainly used as retail premises).[163]

4.103   Certain commonly occurring hereditaments of low rateable value (such as those used for parking, advertisements or electronic communications apparatus) are not permitted relief.[164]

4.104   The amount of relief is different in each case. Post offices with a rateable value below £9,000 are effectively exempt altogether, whereas other hereditaments attract a discount of 50 per cent or less. Again, more generous discounts have been applied on a temporary basis, extended to 31 March 2017.[165]

---

157 LGFA 1988, s43(8A).

158 LGFA 1988, s43(8B).

159 Non-Domestic Rating (Small Business Relief) (Wales) Order 2015.

160 Non-Domestic Rating (Small Business Relief) (Wales) Order 2015, art 6.

161 Non-Domestic Rating (Small Business Relief) (Wales) Order 2015, art 7.

162 Non-Domestic Rating (Small Business Relief) (Wales) Order 2015, art 8.

163 Non-Domestic Rating (Small Business Relief) (Wales) Order 2015, art 9.

164 Non-Domestic Rating (Small Business Relief) (Wales) Order 2015, art 2 definition of 'excepted hereditament'.

165 Non-Domestic Rating (Small Business Relief) (Wales) Order 2015, art 3 and 11–15.

4.105 In Wales mandatory charitable relief 80 per cent takes priority over the small business relief provisions.[166]

4.106 It will be observed that in some cases, in both England and Wales, the effect of the order of priority between different reliefs will mean that a ratepayer benefits from the lower of two levels of relief to which he might have been entitled. In such circumstances it would seem rational for the billing authority to grant discretionary relief up to the level of the greater relief, particularly where the ratepayer is losing out because the hereditament is occupied and used for charitable purposes.

DISCRETIONARY RELIEFS

4.107 The exemptions and reliefs dealt with so far apply automatically when the relevant conditions are satisfied. Billing authorities also have discretionary powers to grant relief in certain circumstances. These are dealt with in turn below. Since 2012, billing authorities also have a general power to grant discretionary relief if it is reasonable to do so having regard to the interests of other council taxpayers in its area.[167]

4.108 In order to apply discretionary relief, the billing authority must first decide that it applies, and then make a determination as to how much relief will be granted.[168] The determination of the amount of any relief may be automatic in accordance with rules set by the billing authority.[169] Discretionary relief will apply on a daily basis. As such a decision that discretionary relief applies must specify the starting date of the relief and may, but need not, specify the end date.[170] A decision to grant relief in respect of any day in a particular financial year must be made no more than six months after the end of that year; any later decision will be automatically invalid.[171] Notice must be given of any decision as to relief and any determination as to the amount of the relief.[172]

---

166 LGFA 1988, s43(8B).
167 LGFA 1988, s47(3) and (5A).
168 LGFA 1988, s47(3) and (1)(a).
169 LGFA 1988, s47(1)(a).
170 Non-Domestic Rating (Discretionary Relief) Regulations 1989, reg 2(1)(a).
171 LGFA 1988, s47(7).
172 Non-Domestic Rating (Discretionary Relief) Regulations 1989, reg 2(1)–(2).

4.109   Where relief is either being withdrawn altogether or reduced, there are stringent notice requirements which must be complied with. At least one year's notice of the change must be given, expiring at the end of a financial year.[173] The combined effect of these two requirements is that notice given on or after 2 April 2017 will not take effect until 31 March 2019. In practice if any decision is communicated along with annual bills (which are usually sent out before 1 April in the relevant year) no such problem will arise. Nevertheless, there are clear advantages to a billing authority in always granting discretionary relief for a fixed period. In that way the relief will expire automatically at the end of the period and either be renewed or not as the authority may decide, without any reference to the notice requirements for revocation of relief.

4.110   In making decisions on whether or not to grant relief (but not to the amount of relief to be granted) a billing authority must have regard to guidance issued by the Secretary of State, in England, or by the Welsh Ministers, in Wales.[174] Somewhat confusingly, the relevant guidance is in both cases entitled *Non-Domestic Rates: Guidance on Rate Relief for Charities and Other Non-Profit Making Organisations*. The English version dates from 2002 and the Welsh from 2004. Much of the material in these publications is a statement of the obvious, but there is an extensive section on the granting of discretionary relief to community amateur sports clubs, which should be studied carefully if a decision is being made in such a case.

4.111   As discretionary relief is in the gift of the billing authority, there is no appeal against the decision as to whether or not to grant it. It is not possible to resist recovery of rates on the basis that discretionary relief should have been granted. However, where the billing authority has made a legal error in its decision to refuse relief (for example, by failing to consider relevant evidence or by ignoring its own policy), it is possible to challenge that decision by making a claim for judicial review.

4.112   A billing authority cannot grant discretionary relief to itself, another billing authority, nor to certain other specified bodies.[175]

---

173 Non-Domestic Rating (Discretionary Relief) Regulations 1989, reg 2(3)–(4).
174 LGFA 1988, s47(5C)–(5D).
175 LGFA 1988, s47(8A). This provision is due to be modified in respect of public conveniences.

Charities and like organisations

4.113   A hereditament which meets the requirement for 80 per cent 'mandatory' charitable relief[176] may also be granted discretionary relief by the billing authority in respect of the final 20 per cent of liability.[177]

4.114   There is a further power to grant discretionary relief to hereditaments which satisfy one of the following two conditions:

    (a)   all or part of the hereditament is occupied for the purposes of one or more institutions or other organisations—

        (i)   none of which is established or conducted for profit, and

        (ii)   each of whose main objects are charitable or are otherwise philanthropic or religious or concerned with education, social welfare, science, literature or the fine arts, or

    (b)   the hereditament—

        (i)   is wholly or mainly used for purposes of recreation, and

        (ii)   all or part of it is occupied for the purposes of a club, society or other organisation not established or conducted for profit.[178]

4.115   Condition (a) does not impose any 'use' requirement; it is enough that the hereditament is *occupied* for the purposes of the organisation. It need not be occupied by the organisation in question, which means that an unincorporated organisation can qualify.[179] Nor is there any requirement that the hereditament be occupied wholly or exclusively for those purposes.[180]

4.116   The organisation may be one whose main purpose is charitable.[181] Once the main purpose or purposes of the organisation have been

---

176   See para 4.85.
177   LGFA 1988, s47.
178   LGFA 1988, s47(5B).
179   *Skegness UDC v Derbyshire Miners Welfare Committee* [1959] AC 807 at p827; *Parker v Ealing Borough Council* [1960] 1 WLR 1398 at p1402.
180   *Royal London Mutual Insurance Society Ltd v Hendon BC* (1958) 3 RRC 76 at p79–80.
181   See para 4.87 on charitable purposes.

established, they must all be charitable in order for the organisation to qualify.[182] There is also provision for other 'main purposes'. The wording of this condition dates in substance from a time when charity law was very restrictive and technical.[183] As such it was necessary to provide for worthy purposes which might fail to meet the strict requirements of charity law.[184] Charitable purposes have now been much more widely defined, so the continued relevance of this part of the condition is questionable (particularly given that billing authorities also have a general power to grant discretionary relief). As such, the many cases decided with reference to this part of the condition under the old law are not included here.

4.117   Condition (b) does incorporate a 'use' requirement. It will be a question of fact whether a hereditament is 'wholly or mainly used' for recreation.[185]

4.118   An element common to both conditions is that the organisation or organisations in question must not be 'established or conducted for profit'. 'For profit' in this phrase means 'for the purpose of making a profit'; an organisation which makes profitable investments in order to further its real purposes accordingly is not established or conducted for profit.[186]

Rural hereditaments

4.119   A billing authority has power to grant further discretionary relief in respect of qualifying hereditaments in rural settlements.[187] This power, like the provisions for mandatory relief in respect of such hereditaments, only applies in England.

---

182 *Berry v St Marylebone Borough Council* [1958] Ch 406.
183 The wording is a relatively unchanged survival from the Rating and Valuation Act 1955, which was the first to put relief for charities on a statutory footing.
184 *National Deposit Friendly Society v Skegness UDC* [1959] AC 293.
185 See the discussion of these terms in the context of mandatory charitable relief at paras 489–495.
186 *National Deposit Friendly Society v Skegness UDC* [1959] AC 293, *Guinness Trust (London Fund) Founded 1890 Registered 1902 v West Ham Borough Council* [1959] 1 WLR 233.
187 LGFA 1988, s47(5A) with reference to s43(6B). See para 4.99 for discussion of the provisions on rural hereditaments.

General discretionary relief

4.120   In all cases not covered by the categories already discussed, the billing authority has a power to grant discretionary relief in any amount. In these cases, however, the billing authority may only grant relief 'if it is satisfied that it would be reasonable for it to do so, having regard to the interests of persons liable to pay council tax set by it'.[188] This requirement has been held in the context of hardship relief[189] to require a 'balancing exercise', in which the claim to relief is balanced against the interests of council taxpayers.[190] It has been suggested that, without such a requirement, it would be unlawful to rely on the need to maintain the billing authority's revenue when determining whether or not to grant relief.[191] This argument may gain strength from the fact that the requirement to consider the interests of other council taxpayers is the only thing which distinguishes the general power to grant discretionary relief from the specific powers set out above, which tends to suggest that the requirement is supposed to add something which would not otherwise be there.

UNOCCUPIED HEREDITAMENTS

4.121   There are, broadly speaking, two sources of exemptions and reliefs for unoccupied hereditaments. First, certain hereditaments are excluded from the scope of unoccupied rating altogether by the terms of the relevant regulations.[192] There are several such exceptions. The most common are the first two, which relate to the amount of time the hereditament has been unoccupied for. These two provisions have already been discussed.[193] The others are dealt with in turn below.

---

188 LGFA 1988, s47(5A).
189 See para 4.145 onwards.
190 *R v Birmingham ex p Mushtaq* [1995] RVR 142 at p144.
191 *R v Liverpool CC ex p Windsor Securities Ltd* [1979] RA 159 at p181 per Lawton LJ. See also *R v Tower Hamlets BC ex p Chetnik Developments Ltd* [1989] AC 858 at p879. In the latter case the context was one of refunds of overpayments, which more obviously tends to exclude the interests of other taxpayers as a consideration.
192 The Non-Domestic Rating (Unoccupied Property) (England) Regulations 2008 and The Non-Domestic Rating (Unoccupied Property) (Wales) Regulations 2008.
193 See para 3.22 in Chapter 3 on liability.

4.122   Second, there are provisions in LGFA 1988 relating to charities and like organisations and to discretionary relief which may apply to unoccupied hereditaments which would otherwise give rise to liability.

Occupation prohibited by law or action of public authority

4.123   A hereditament will be exempt from unoccupied rates if its owner is prohibited by law from occupying it or allowing it to be occupied,[194] or if it is kept vacant by reason of action taken by a public authority with a view to prohibiting occupation of it or to acquiring it.[195]

4.124   In order to secure this exemption, any use of the hereditament must be prohibited, not just its use for the purposes indicated by its description in the list.[196] There must also be an actual prohibition, whether this is imposed by statute[197] or by a notice served under a statute.[198] A good example is that of planning permission. The absence of planning permission will render a use unlawful in general terms, but it is not prohibited for these purposes unless enforcement action has been taken.[199] The action taken under the statute must target occupation directly[200] or by necessary implication.[201]

---

194 The Non-Domestic Rating (Unoccupied Property) (England) Regulations 2008 and The Non-Domestic Rating (Unoccupied Property) (Wales) Regulations 2008, reg 4(c).

195 The Non-Domestic Rating (Unoccupied Property) (England) Regulations 2008 and The Non-Domestic Rating (Unoccupied Property) (Wales) Regulations 2008, reg 4(d).

196 *Hailbury Investments Ltd v Westminster CC* [1986] 1 WLR 1232.

197 See, for example, *Tower Hamlets LBC v St Katherine-by-the-Tower Ltd* [1982] RA 261 in which the relevant statutory provision stated that '[n]o building . . . shall be occupied . . . until the council have issued a certificate' in respect of a fire escape. Occupation was thus prohibited by the statute and no notice was required.

198 Contrast *Westminster CC v Regent Lion Properties Ltd* [1990] RA 121 at p131, where a notice had been served under health and safety legislation, with *Pall Mall Investments (London) Ltd v Gloucester CC* [2014] EWHC 2247 (Admin), where no such notice had been served. Occupation was prohibited in the first case but not the second.

199 *Westminster CC v Regent Lion Properties Ltd* [1990] RA 121.

200 In *Henderson v Liverpool MDC* [1980] RA 238, a condition to a planning permission required demolition. This was held not to be a prohibition on occupation in the meantime.

201 In *Westminster CC v Regent Lion Properties Ltd* [1990] RA 121 a notice prevented the refurbishment work which would have been necessary before occupation could take place until certain remedial works had been done. Occupation was held to be prohibited in these circumstances.

Listed buildings and scheduled ancient monuments

4.125 A hereditament which is the subject of a building preservation notice or is included in list compiled under section 1 of the
Planning (Listed Buildings and Conservation Areas) Act 1990
(in other words, a listed building) is exempt from unoccupied
rates.[202] So is a hereditament included in the schedule of monuments compiled under section 1 of the Ancient Monuments and
Archaeological Areas Act 1979 (in other words, a scheduled
ancient monument).[203]

4.126 It will be remembered that unoccupied rates only apply in the
first place to 'relevant' hereditaments; i.e. hereditaments consisting of a building or part of a building together with land ordinarily
used for the purposes of the building or part.[204] Issues can arise in
respect of this exemption, therefore, where the hereditament does
not correspond exactly to the listed building or scheduled ancient
monument.

4.127 The courts have confirmed that it is not enough, to secure exemption, for a part of the hereditament to be a listed building.[205]
The whole hereditament must be listed. It is therefore very important to identify the extent of the listed building. This may be more
extensive than first appears, because of the effect of the relevant
statutory provisions defining a listed building. These provide as
follows:

> (5) In this Act 'listed building' means a building which is for the
> time being included in a list compiled or approved by the
> Secretary of State under this section; and for the purposes of
> this Act—

---

202 The Non-Domestic Rating (Unoccupied Property) (England) Regulations 2008
and The Non-Domestic Rating (Unoccupied Property) (Wales) Regulations
2008, reg 4(e).

203 The Non-Domestic Rating (Unoccupied Property) (England) Regulations 2008
and The Non-Domestic Rating (Unoccupied Property) (Wales) Regulations
2008, reg 4(f).

204 The Non-Domestic Rating (Unoccupied Property) (England) Regulations 2008
and The Non-Domestic Rating (Unoccupied Property) (Wales) Regulations
2008, reg 2 and 3.

205 *Providence Properties Ltd v Liverpool* [1980] RA 189, approved by the House of
Lords in *Debenhams v Westminster* [1987] AC 396. Applied in respect of LGFA
1988 by *Richardson Developments Ltd v Birmingham CC* (1999, unreported).

(a) any object or structure fixed to the building;

(b) any object or structure within the curtilage of the building which, although not fixed to the building, forms part of the land and has done so since before 1st July 1948, shall, subject to subsection (5A)(a), be treated as part of the building.

(5A) In a list compiled or approved under this section, an entry for a building situated in England may provide—

(a) that an object or structure mentioned in subsection (5)(a) or (b) is not to be treated as part of the building for the purposes of this Act.[206]

4.128 Thus curtilage structures dating from before 1 July 1948 will automatically form part of a listed building, unless the list states otherwise. 'Objects or structures' fixed to the building will also automatically form part of a listed building, unless the list states otherwise. The House of Lords has held that 'the word "structure" is intended to convey a limitation to such structures as are ancillary to the listed building itself, for example the stable block of a mansion house, or the steading of a farmhouse, either fixed to the main building or within its curtilage . . . the concept envisaged is that of principal and accessory'.[207] The question of whether a 'structure' is ancillary to a listed building and therefore forming part of it is to be resolved 'objectively' rather than by reference to the use made of them by a particular occupier'.[208] These provisions accordingly will not apply where the 'structure' is an independent building,[209] or where the listed part is in fact ancillary to the 'structure'.[210]

4.129 The requirement that the whole hereditament be a listed building is subject to the proviso that a hereditament may include 'land ordinarily

---

206 Planning (Listed Buildings and Conservation Areas) Act 1990, s1.

207 *Debenhams v Westminster* [1987] AC 396 at p403.

208 *Debenhams v Westminster* [1987] AC 396 at p404.

209 As was the case in *Debenhams v Westminster* [1987] AC 396, which concerned a hereditament composed of two buildings linked by a bridge and tunnel across a street. One of the buildings was listed, the other was not. The House of Lords held that the linked building was not a 'structure fixed' to the listed building.

210 As was the case in *Richardson Developments Ltd v Birmingham CC* (1999, unreported).

used for the purposes of the building or part'. Such land cannot form part of a listed building but does not prevent the hereditament as a whole from attracting the exemption.[211]

4.130 There is no requirement for the hereditament to be named in the list of listed buildings. As such, where the hereditament forms part only of a listed building, it will still attract the exemption.[212]

4.131 There have been no reported cases dealing with buildings the subject of a building preservation notice or scheduled ancient monuments.

4.132 Building preservation notices are a form of emergency protection for as-yet-unlisted buildings of special architectural or historic interest which are at risk of demolition or alteration. Whilst such a notice is in force, it is as if the building were listed.[213] The same approach to exemption from unoccupied rates should therefore be applied to them as is applied to listed buildings.[214]

4.133 The provisions on scheduled ancient monuments are similar to those relating to listed buildings. As such, it would again seem that the whole hereditament must form part of a scheduled ancient monument in order to attract exemption. The monument and its site can form part of the scheduling,[215] which is accordingly to be defined strictly with reference to the schedule plan.[216] This means that there is no scope to include ancillary structures by implication. Furthermore, there would appear to be no scope for the hereditament to include associated land which is not subject to the scheduling. The scheduling itself *can* include such land so, unlike in the case of listed buildings, there is no need to widen the scope of the exemption by implication to include it.

---

211 *Debenhams v Westminster* [1987] AC 396 at p404–405. The House of Lords was referring to the previous statutory provisions, which contained a slightly different provision on associated land, but the logic is the same in respect of the current provisions.

212 *Ge Bowra Group Ltd v Thanet DC* [2007] EWHC 1077 (Admin).

213 Planning (Listed Buildings and Conservation Areas) Act 1990, s3(5).

214 *Debenhams v Westminster* [1987] AC 396 at p404. Lord Keith indicated that the literal words could cover a situation where only part of a hereditament was subject to a preservation notice, but considered that it was unlikely that Parliament had intended to treat listing and building preservation notices differently in this respect.

215 The Ancient Monuments and Archaeological Areas Act 1979, ss1 and 61.

216 *R v Bovis Construction Ltd* [1994] Crim LR 938.

Low rateable value

4.134 Hereditaments with a rateable value below a specified level are
       exempt from unoccupied rates.[217] The precise amount has varied over
       the years, and is currently set at £2,600.

Deceased and insolvent owners

4.135 The following categories of hereditaments are exempt from unoccu-
       pied rates:[218]

   (h) whose owner is entitled to possession only in his capacity as
        the personal representative of a deceased person;
   (i) where, in respect of the owner's estate, there subsists a bank-
        ruptcy order within the meaning of section 381(2) of the
        Insolvency Act 1986;
   (k) whose owner is a company which is subject to a winding-up
        order made under the Insolvency Act 1986 or which is being
        wound up voluntarily under that Act;
   (l) whose owner is a company in administration within the
        meaning of paragraph 1 of Schedule B1 to the Insolvency
        Act 1986 or is subject to an administration order made under
        the former administration provisions within the meaning of
        article 3 of the Enterprise Act 2002 (Commencement No. 4
        and Transitional Provisions and Savings) Order 2003 4;
   (m) whose owner is entitled to possession of the hereditament in
        his capacity as liquidator by virtue of an order made under
        section 112 or section 145 of the Insolvency Act 1986.

4.136 Sub-paragraph (k) has given rise to a widespread avoidance scheme.
       Unoccupied premises were leased to companies created specifically
       for the purpose at a nominal rent. The companies were then placed
       into members' voluntary liquidation such that the premises became

---

217 The Non-Domestic Rating (Unoccupied Property) (England) Regulations 2008
    and The Non-Domestic Rating (Unoccupied Property) (Wales) Regulations
    2008, reg 4(g).
218 The Non-Domestic Rating (Unoccupied Property) (England) Regulations 2008
    and The Non-Domestic Rating (Unoccupied Property) (Wales) Regulations
    2008, reg 4. The letters refer to sub-paragraphs in this regulation.

exempt from unoccupied rates. The liquidators were then removed and not replaced meaning that the liquidation was prolonged and the exemption continued until the freehold owner had need of the premises. The company facilitating this scheme was wound up on the basis that it was 'just and equitable' to do so.[219]

Charities/sports clubs

4.137 An unoccupied hereditament will be 'zero rated' in respect of any day on which: (a) it is owned by a charity or trustees of a charity; and (b) it appears that when next in use the hereditament will be wholly or mainly used for charitable purposes (whether of that charity or of that and other charities).[220] There is a similar provision for registered community amateur sports clubs.[221] These tests share various similarities with the tests for charitable relief in respect of occupied hereditaments (particularly the requirement of 'wholly or mainly used'), and reference should be made to the discussion of the tests for occupied hereditaments accordingly.

4.138 In many cases the application of this section will be straightforward. There are however three finer points of interpretation to note, which will become relevant in more complex cases.

4.139 The first relates to the purposes for which the hereditament must be used when next in use. The High Court has held that the words in the Act mean, in effect, 'used for the charitable objects of the owning charity, accompanied or not by other charitable purposes'.[222] This means that it does not need to be shown by the ratepayer that it will be the next user. However, it does need to be shown that the next user will have the same charitable purposes as the owner, whether or not it also has other charitable purposes. It is to be noted that the position for community amateur sports clubs is as yet unresolved and may be different; the provisions on these clubs appear to lay more stress on the next use being related to 'that club' which is the owner.

---

219 *PAG Management Services Ltd* [2015] EWHC 2404 (Ch).
220 LGFA 1988, s45A(2).
221 LGFA 1988, s45A(3).
222 *Preston City Council v Oyston Angel Charity* [2012] EWHC 2005 (Admin); [2012] RA 357 at [38].

4.140   The second relates to when the conditions must be met. The High
        Court has described the statutory test as a 'forward looking' exercise.[223]
        This is apparently supported by the following features of the statutory
        scheme. Zero rating is to be applied on any day in respect of which
        the conditions are satisfied. The second condition is that 'it appears'
        that when next in use the use will be wholly or mainly charitable.
        This suggests that the conditions must be shown to be met as things
        stood on the day for which zero rating is sought, rather than, for
        example, the day on which the matter is considered. In other words,
        it is apparently not permissible to use hindsight in resolving the ques-
        tion of whether zero rating should be applied. This point, which has
        not been tested in the higher courts, will be very relevant in a situa-
        tion where a billing authority has refused to apply zero rating because
        it does not appear that the next use will satisfy the relevant test, but
        then in due course the next use does in fact amount to a use which is
        wholly or mainly for charitable purposes.

4.141   The third is about what is meant by the expression 'when next in
        use'. Does it relate simply to the next use of the hereditament, or to
        the next period of occupation? The latter would seem more logical as
        it would provide some degree of correlation with the provisions on
        charitable relief for occupied premises, but the former appears to be
        required by the literal words of the provision. If the former is correct,
        there is also uncertainty about the period within which it must appear
        that the hereditament will be wholly or mainly used for charitable
        purposes – must it be so used at the first instant of the next use, over
        the whole period of the use or somewhere in between? If, on the other
        hand, the reference is really to the next occupation then, given that
        this is assessed on a daily basis, it would appear that the first day of
        occupation will be the relevant period. These questions may have to
        be resolved in the higher courts if an appropriate case presents itself.

Discretionary relief

4.142   The billing authority's general power to grant discretionary relief,
        taking account of the interests of its council taxpayers, applies to
        unoccupied hereditaments as it does to occupied ones.[224]

223 *Preston City Council v Oyston Angel Charity* [2012] EWHC 2005 (Admin); [2012]
     RA 357 at [19], [27], [32].
224 See para 4.120.

4.143   The specific powers to grant discretionary relief in respect of charities and like institutions (and, in England, certain rural hereditaments) also apply to unoccupied hereditaments.[225] The question of whether a hereditament is occupied or used for a particular purpose is to be resolved in this case by looking at the apparent position when it is next in use.[226]

4.144   A billing authority cannot grant discretionary relief from unoccupied rates where it appears that, when next in use, the unoccupied hereditament will be occupied by a billing authority or certain other public bodies.[227]

HARDSHIP RELIEF

4.145   A billing authority has a further power to grant total or partial relief in any case of either occupied or unoccupied rates, provided that it is satisfied of two things:

   (a)   the ratepayer would sustain hardship if the authority did not do so; and
   (b)   it is reasonable for the authority to do so, having regard to the interests of persons liable to pay council tax set by it.[228]

4.146   'Hardship' is different from poverty; it can accordingly be suffered even by a limited company or other corporate ratepayer. It will also be relevant, when considering a corporate ratepayer, to consider any possible hardship to the shareholders or other individuals behind the ratepayer.[229]

4.147   The Court of Appeal has said that whether or not the ratepayer would suffer hardship is to be resolved 'in the light of common sense', which appears to imply that there can be no further sensible definition of the term.[230] The hardship in question is hardship to the ratepayer. It is therefore very important for the ratepayer to provide information

---

225   See paras 4.113–118.
226   LGFA 1988, s48(3) and (5).
227   LGFA 1988, s48(4) and s47(8A).
228   LGFA 1988, s49(2).
229   *R v Liverpool ex p Caplin* [1984] RVR 132 at p134; *Wakefield MDC v Huzminor Investment Developments Ltd* [1989] RVR 108 at p110.
230   *R v Liverpool CC ex p Windsor Securities Ltd* [1979] RA 159 at p177–178.

to the billing authority about its financial position, because hardship means something more than substantial detriment. Hardship is a matter of degree, and the extent of hardship will be closely related to the amount of relief that a billing authority may decide to grant.[231] The billing authority is not required to take into account hardship which arises because of the improvidence of the ratepayer.[232]

4.148   The billing authority must also consider the interests of other council taxpayers. This requirement has been held to require a 'balancing exercise', in which the hardship to one ratepayer is balanced against the interests of council taxpayers.[233] How that balancing exercise is to be carried out is a matter for the billing authority, which must somehow decide how these competing and very different considerations are to be weighed against each other. In some cases, of course, it may be that granting hardship relief will overall be in the interests of other council taxpayers – if, for example, it would prevent a local employer going out of business.

---

231 *R v Liverpool CC ex p Windsor Securities Ltd* [1979] RA 159 at p177–178, p183.

232 *R v Liverpool ex p Caplin* [1984] RVR 132 at p133–134.

233 *R v Birmingham ex p Mushtaq* [1995] RVR 142 at p144.

5

Collection of rates

5.1 The procedure for collection of rates by billing authorities is set out in the NDR Collection Regulations 1989.[1] These provide for the service of demand notices, which are commonly referred to simply as bills. If an amount set out in a demand notice is not paid, then in theory there are several different methods of enforcement available to the billing authority. In practice, however, 99 per cent of unpaid liabilities are enforced by seeking a liability order in the Magistrates' Court.

5.2 There is a parallel procedure for the collection of rates by the Secretary of State in respect of central list hereditaments.[2] This is dealt with separately below.

DEMAND NOTICES

Requirement for demand notices

5.3 The first step in collection is the service of a demand notice; it is the service of such a notice that crystallises the ratepayer's liability to rates into a concrete obligation to pay.[3] The billing authority must serve such a notice on each ratepayer for each financial year.[4] A notice may relate to more than one hereditament. In England, but not in Wales,

---

1 Non-Domestic Rating (Collection and Enforcement) (Local List) Regulations 1989.
2 Under the Non-Domestic Rating (Collection and Enforcement) (Central List) Regulations 1989.
3 Non-Domestic Rating (Collection and Enforcement) (Local List) Regulations 1989, reg 7(6).
4 Non-Domestic Rating (Collection and Enforcement) (Local List) Regulations 1989, reg 4.

it may relate to more than one financial year, whereby a ratepayer is being billed for the current year at the same time as he is billed for previous years. In Wales, therefore, if a demand notice deals with more than one financial year it cannot be relied upon to recover the rates demanded in it.[5]

Timing of demand notices

5.4   Demand notices must be served 'on or as soon as practicable after' 1 April or the first day in the financial year on which liability arises. They can be served before the beginning of the financial year on the person who then appears to be liable.[6] The word 'practicable' in this context means 'feasible' and 'possible to be accomplished with known means and known resources'.[7] Local authority resources will thus be relevant to what is 'practicable', but a 'billing authority will not be able to rely upon the suggestion that home-grown problems and inefficiencies rendered impracticable what would otherwise have been practicable'.[8] The issue in each case is by what point it was feasible for the billing authority to have discovered the identity of the ratepayer.[9] An alteration to the list will not excuse earlier failures in respect of the same hereditaments.[10] However, a billing authority's failure to notify the valuation officer of information relevant to the alteration of the list will not invalidate a demand notice served after the list has eventually been altered, even if a ratepayer is prejudiced by that failure.[11]

5.5   This time limit is important, because if it is not complied with the demands may not be enforceable. A ratepayer must show that he has suffered some procedural or substantive prejudice as a result of the late

---

5   *R (JJB Sports plc) v Telford and Wrekin BC* [2008] EWHC 2870 (Admin), in which the district judge in the Magistrates' Court described the issuing of 'multi-bills' dealing with multiple years as 'disgraceful' and 'an abuse of the court process', without any apparent disagreement from the High Court.

6   Non-Domestic Rating (Collection and Enforcement) (Local List) Regulations 1989, reg 5.

7   *North Somerset DC v Honda Motor Europe Ltd and others* [2010] EWHC 1505 at [63]–[64].

8   *North Somerset DC v Honda Motor Europe Ltd and others* [2010] EWHC 1505 at [64].

9   *Encon Insulation Ltd v Nottingham CC* [1999] RA 382.

10   *North Somerset DC v Honda Motor Europe Ltd and others* [2010] EWHC 1505 at [69].

11   *R (Secerno Ltd) v Oxford Magistrates' Court* [2011] EWHC 1009 (Admin).

service of the notices.[12] If he can show this, then it will constitute a valid defence to recovery of the money whether in the Magistrates' Court or elsewhere.[13]

Contents of demand notices

5.6   The contents of a demand notice must accord with the provisions of the Council Tax and Non-Domestic Rating (Demand Notices) (England) Regulations 2003 or, in Wales, the Non-Domestic Rating (Demand Notices) (Wales) Regulations 1993. Failure to adhere to these regulations, however, need not result in the invalidity of the notice. Even if it does, there is specific provision in the regulations to the effect that where such failure was a mistake, the amounts will continue to be payable.[14]

5.7   A demand notice served in respect of future periods (as almost all demand notices are) must require payment by instalments unless the ratepayer and the billing authority have agreed otherwise.[15] The number of instalments is 10, or one less than the number of months remaining in the relevant year after the issue of the notice.[16] In England the ratepayer can serve an 'instalment notice' on the billing authority requiring payment to be demanded in 12 instalments, or the same number as there are months remaining in the relevant year.[17]

---

12 *R (LB Waltham Forest) v Waltham Forest Magistrates' Court and Yem Yom Ventures Ltd* [2008] EWHC 3579 (Admin) at [45], *R (JJB Sports plc) v Telford and Wrekin BC* [2008] EWHC 2870 (Admin), *North Somerset DC v Honda Motor Europe Ltd and others* [2010] EWHC 1505 at [60]. The suggestion in *Encon Insulation Ltd v Nottingham CC* [1999] RA 382 that late service would automatically lead to invalidity has been disapproved by these cases.

13 *R (LB Waltham Forest) v Waltham Forest Magistrates' Court and Yem Yom Ventures Ltd* [2008] EWHC 3579 (Admin) at [46], *North Somerset DC v Honda Motor Europe Ltd and others* [2010] EWHC 1505 at [34].

14 Council Tax and Non-Domestic Rating (Demand Notices) (England) Regulations 2003, reg 4(2); Non-Domestic Rating (Demand Notices) (Wales) Regulations 1993, reg 5(1).

15 Non-Domestic Rating (Collection and Enforcement) (Local List) Regulations 1989, reg 7(1).

16 Non-Domestic Rating (Collection and Enforcement) (Local List) Regulations 1989, sch 1 para 1.

17 Non-Domestic Rating (Collection and Enforcement) (Local List) Regulations 1989, reg 7(1E).

5.8   If instalments are not paid, a further notice is to be served demanding that they be paid.[18] If that notice is not complied with, or if further instalments are subsequently missed, the whole year's liability becomes payable as a lump sum.[19]

5.9   By contrast, a notice issued in respect of past periods must require payment as a lump sum. At least 14 days must be given for payment.[20]

5.10 These provisions on what a demand notice must contain and the terms it must set for payment do not necessarily constrain a billing authority's freedom of action. A billing authority does not have to insist on payment according to the terms of the demand notice, and can come to an arrangement with the ratepayer for payment in a different manner. Any such agreement rests purely on the billing authority's discretion not to take further steps of enforcement, however, and the ratepayer remains vulnerable to such further steps unless he pays in accordance with the terms of the demand notice.

5.11 It frequently happens that the liability is less than or more than was estimated when the demand notice was issued at or before the beginning of the relevant year. In any such case the billing authority is obliged to serve a further notice stating the revised amount and adjusting the amounts payable accordingly. If there is a further liability, this is due as a lump sum. If there has been an overpayment, this must either be repaid if the ratepayer so requires or otherwise repaid or credited to any subsequent liability of the ratepayer.[21]

Service of demand notices

5.12 A failure to serve a demand notice correctly may mean that a court is prevented from ordering repayment of the sums demanded and/or may lead to a court order being set aside.[22] It is therefore very important for billing authorities to ensure that demand notices are served correctly.

---

18 Non-Domestic Rating (Collection and Enforcement) (Local List) Regulations 1989, reg 8(1).

19 Non-Domestic Rating (Collection and Enforcement) (Local List) Regulations 1989, reg 8(2).

20 Non-Domestic Rating (Collection and Enforcement) (Local List) Regulations 1989, reg 7(5).

21 Non-Domestic Rating (Collection and Enforcement) (Local List) Regulations 1989, reg 9. See Chapter 7 on securing repayment in contested cases.

22 On which, see para 5.61 onwards.

5.13   The service of notices by billing authorities is governed by the Local Government Act 1972, section 233. This requires a notice to be served in one of two ways. The first is by delivering it to the recipient. In order to be effective this method requires proof of actual receipt by the recipient. The second, more convenient method is by delivering the notice to the recipient's 'proper address'. It can either be left there or, more normally, posted.

5.14   If a notice is served at the 'proper address' various presumptions operate. If it is left at the address by a council officer, it is deemed to have been served whether or not it is ever received by the ratepayer.[23] If it is (correctly) posted then it is presumed to have been served on the day when it would have arrived in the normal course of post.[24] The fact of service cannot be disputed.[25] However, where service has to be effected by a certain time, it is open to the recipient to prove that the notice was not received in the normal course of post or at all, and thus to show that it was not served in time.[26] In the context of a dispute over non-domestic rates the time of service would not seem to be particularly important in most cases, unless there is an allegation that the billing authority failed to serve the notice 'as soon as practicable'. As such, if notices are correctly posted to the 'proper address', service cannot normally be disputed.

5.15   The 'proper address' of a person for these purposes is his last known address. For a corporate body, on the other hand, the proper address is the address of its registered office or (for statutory corporations and foreign corporations) its principal office in the United Kingdom. Notices can be served on the clerk or secretary of a corporate body and, in the case of a partnership, either on a partner or a person having control or management of the partnership's business. A ratepayer can specify a different address to be used for service, which will be treated as his proper address.[27]

5.16   Further flexibility is introduced in terms of the address at which demand notices must be served by the NDR Collection Regulations 1989.

23 *Rushmoor BC v Reynolds* (1991) 23 HLR 495.
24 This is the effect of s7 of the Interpretation Act 1978.
25 See *A/S Cathrineholm v Norequipment Trading Ltd* [1972] 2 QB 314 per Denning LJ at p322.
26 See, e.g., *Hewitt v Leicester Corp* [1961] 1 WLR 855, *Maltglade Ltd v St Albans RDC* [1972] 1 WLR 1230.
27 Local Government Act 1972, s233(2)–(5).

Where the notice relates to at least one hereditament which is a place of business of the ratepayer, the notice can either be left at or posted to that address.[28] Premises let out by a landlord are not normally a 'place of business' of the landlord, unless it is shown that he physically attends there to collect the rent.[29] This provision will accordingly be of limited use to a billing authority seeking to recover unoccupied rates.

5.17   Options for service by electronic means are also provided by the NDR Collection Regulations 1989. The demand notice itself can be served by electronic communication when an address has been notified to the billing authority for that purpose. Alternatively, the billing authority can simply provide the ratepayer with notification that the demand notice is available on a website if this method is agreed with the ratepayer.[30] In addition, the information required to be contained in a demand notice can be supplied to the ratepayer by being made available on a website even without the ratepayer's agreement, as long as the demand notice itself informs the ratepayer that it is available.[31] If a hard copy of the information is requested, it must be supplied.[32]

COMMENCING MAGISTRATES' COURT PROCEEDINGS

Reminder notices

5.18   Before commencing proceedings for a liability order in the Magistrates' Court to recover sums due under a demand notice, the billing authority must serve a reminder notice allowing seven more days to pay the amount due.[33] Reminder notices are sometimes referred to, incorrectly, as 'final notices', perhaps because this is the name of the equivalent

---

28 Non-Domestic Rating (Collection and Enforcement) (Local List) Regulations 1989, reg 2(2).
29 *Chowdhury v Westminster CC* [2013] EWHC 1921 (Admin) at [31].
30 Non-Domestic Rating (Collection and Enforcement) (Local List) Regulations 1989, reg 2(3).
31 Non-Domestic Rating (Collection and Enforcement) (Local List) Regulations 1989, reg 2(3A) and (8).
32 Non-Domestic Rating (Collection and Enforcement) (Local List) Regulations 1989, reg 2(3B) and (9).
33 Non-Domestic Rating (Collection and Enforcement) (Local List) Regulations 1989, reg 11(1) and 12(1).

notice for council tax purposes.[34] There is no requirement to serve such a notice where notice has already been given of a failure to pay instalments, even if the proceedings are to recover the whole liability for that year and not just the missed instalments.[35]

5.19 A reminder notice must be served on the ratepayer in the same way as a demand notice.[36]

Complaint and summons

5.20 The actual proceedings in the Magistrates' Court are to be commenced by 'making complaint . . . requesting the issue of a summons'.[37] This is the standard method of instituting civil proceedings in the Magistrates' Court. In theory the complaint is made to the court by the billing authority and then a summons is issued by the court. In practice the two are usually generated as one document which is simply rubber-stamped by the court before being sent out by the billing authority. The billing authority makes the complaint and so is known as the 'Complainant' in the court proceedings; the ratepayer is known as the 'Defendant'.

5.21 The summons must be served on the ratepayer. This is a vital step, because if the summons is not served correctly, the court will have no power to make a liability order.[38] A summons may be served on the ratepayer:

    (a) by delivering it to him;

    (b) by leaving it at his usual or last known place of abode, or in the case of a company, at its registered office;

    (c) by sending it by post to him at his usual or last known place of abode, or in the case of a company, to its registered office;

    (d) where all or part of the sum to which it relates is payable with respect to a hereditament which is a place of business of

34 As to which, see para 10.13.

35 Non-Domestic Rating (Collection and Enforcement) (Local List) Regulations 1989, reg 11(3) and 12(1).

36 On which, see 5.12 onwards.

37 Non-Domestic Rating (Collection and Enforcement) (Local List) Regulations 1989, reg 12(2).

38 *Chowdhury v Westminster CC* [2013] EWHC 1921 (Admin) at [28].

the person, by leaving it at, or by sending it by post to him at, the place of business; or

(e) by leaving it at, or by sending it by post to him at, an address given by the person as an address at which service of the summons will be accepted.[39]

5.22 Where the summons is left at an address or sent by post, the deeming provisions discussed above will apply.[40] No liability order can be made unless 14 days have elapsed since the summons was served.[41] As such it would seem that the time of service is of importance and that, as a consequence, it is open to a ratepayer to prove that a summons deemed served in the ordinary course of post has not in fact been received by him.

5.23 The limitation period for bringing liability order proceedings is '6 years beginning with the day on which [the sum] became due'. A sum does not become due until it has been the subject of a demand notice.

HEARINGS IN THE MAGISTRATES' COURT

5.24 In considering whether to make a liability order, the Magistrates' Court 'shall make the order if it is satisfied that the sum has become payable by the defendant and has not been paid'.[42]

5.25 The majority of liability order hearings in the Magistrates' Court are uncontested, or are resolved by a very simple process of consideration. The billing authority will usually not be legally represented but will have authorised an officer to conduct the liability order hearings for it, as it is entitled to do.[43] The majority of ratepayers do not attend, and the court is able to proceed in their absence provided that it is proved on oath that the summons was served on them a reasonable time before the hearing.[44] In practice, tens or hundreds of liability orders

---

39 Non-Domestic Rating (Collection and Enforcement) (Local List) Regulations 1989, reg 13(2).

40 See para 5.14.

41 Non-Domestic Rating (Collection and Enforcement) (Local List) Regulations 1989, reg 13(2A).

42 Non-Domestic Rating (Collection and Enforcement) (Local List) Regulations 1989, reg 12(5).

43 Local Government Act 1972, s223.

44 Magistrates' Court Act 1980, s55.

may be issued in one go relying on this provision (as part of a so-called 'bulk list'). There will rarely be any formal 'order' issued by the court; the practice is often for an officer of the court or a magistrate to sign a list prepared by the billing authority. This practice, although rather unusual in court proceedings, is not unlawful.[45]

5.26   If there is a serious dispute about liability, this should be notified by the ratepayer to the billing authority. Where the ratepayer has indicated that it intends to contest liability, the normal practice is for the case to be adjourned from the 'bulk list' by agreement. It should preferably be adjourned to a hearing before a district judge at which point directions can be given by the judge for evidence and skeleton arguments to be exchanged before a final hearing of the case. If a billing authority fails to make such an arrangement it is likely to be penalised in costs for requiring the ratepayer to attend the bulk list hearing unless the court and both parties are prepared for the issues to be dealt with there and then.[46]

5.27   Following any directions hearing, the case will be listed for a final hearing, preferably before a district judge rather than a bench of three lay magistrates. In terms of procedure, such a hearing is a civil case in the Magistrates' Court, and is therefore governed by the somewhat archaic Magistrates' Court Rules 1981. In terms of evidence, these provide for a three-stage process whereby the billing authority calls its evidence, the ratepayer calls his evidence, and the billing authority is then provided with a further opportunity to call rebuttal evidence.[47] In practice this procedure is seldom followed. Generally, the billing authority will give its evidence, followed by the ratepayer. Similarly, the rules provide for the billing authority to make an opening speech and for the defendant to make either an opening or closing speech.[48] This can put the billing authority at a significant disadvantage, because it loses the ability to make a closing speech once all the evidence has been heard and cross-examined. As such, a well-advised billing authority will generally seek the court's permission to make a

---

45 *Georgiou v Redbridge LBC* [2013] EWHC 4579 (Ch) at [6]–[7].
46 *R (on the application of North East Lincolnshire DC) v Grimsby and Cleethorpes Magistrates' Court* [2013] EWHC 2368 (Admin) is an example of the difficulties that can arise when a billing authority refuses to agree an adjournment, but equally is not prepared to deal with the substantive dispute on the day of the bulk list hearing.
47 Magistrates' Court Rules 1981, r14(1)–(3).
48 Magistrates' Court Rules 1981, r14(1), (2), (4).

second speech. As Complainant, it will therefore have the first and the last word in terms of making submissions.[49]

5.28 The burden of proof is on the billing authority to establish that a liability order should be made. However, in terms of the evidence that must be provided to do that, there is what has been described, rather unhappily, as a 'swinging' burden of proof. The billing authority may issue a summons when it has reasonable grounds to believe that a ratepayer is liable. From that point, the evidential burden works as follows:

> all the [billing] authority has to show in the first instance is that (a) the rate in question has been duly made and published; (b) it has been duly demanded from the respondent, and (c) it has not been paid. If these three things are shown, the burden then falls on the respondent to show sufficient cause for not having paid the sum demanded.[50]

5.29 What the ratepayer must do to show that he is not liable will depend on all the circumstances. It may be that in some circumstances he need do little more than say 'I am not the rateable occupier'.[51] On the other hand, merely producing a purported lease with no supporting evidence may not be enough to discharge the evidential burden on the ratepayer.[52]

5.30 The Magistrates' Court in a liability order application cannot question the contents of the rating list,[53] and the contents of that list can be proved by the production of a copy purporting to be certified by the billing authority.[54]

5.31 As noted above, the Magistrates' Court 'shall' make the order if it is satisfied that the sum has become payable but has not been paid.[55]

49 Magistrates' Court Rules 1981, r14(5)–(6).
50 *Ratford v Northavon DC* [1987] QB 357. This was a decision on the previous provisions but the reasoning still applies: *Pall Mall Investments v LB Camden* [2013] EWHC 459 (Admin) at [16].
51 *Forsythe v Rawlinson* [1981] RVR 97 at p98.
52 See *Pall Mall Investments v LB Camden* [2013] EWHC 459 (Admin).
53 Non-Domestic Rating (Collection and Enforcement) (Local List) Regulations 1989, reg 23(1) and *R (Vtesse Networks Ltd) v North West Wiltshire Magistrates' Court* [2009] EWHC 3283 (Admin).
54 Non-Domestic Rating (Collection and Enforcement) (Local List) Regulations 1989, reg 23(2).
55 Non-Domestic Rating (Collection and Enforcement) (Local List) Regulations 1989, reg 12(5).

It is unclear whether this allows a ratepayer to raise as a defence the fact that he has overpaid in respect of previous years or other liabilities. The use of the mandatory word 'shall' suggests that the court has no discretion. However, it might well be arguable that a sum demanded has not in fact 'become payable' if it can be offset against a right to repayment of other sums paid previously to the billing authority. Alternatively, an entitlement to a repayment might enable a ratepayer to 'show why he has not paid the sum outstanding'.[56] An authority dealing with the very similar council tax provisions decides that 'the Magistrates' Court must enquire into questions as to whether the taxpayer is entitled to set off monies owed by the billing authority'.[57]

5.32   If a billing authority acknowledges the right to a credit or repayment but nevertheless seeks to recover the full sum for a later year, it may well be acting in bad faith. Cases decided under the previous provisions support the argument that a Magistrates' Court may refuse to make an order in such circumstances.[58]

5.33   If a billing authority does not acknowledge the right to a repayment, the raising of this defence may require the court to investigate and make rulings on the ratepayer's entitlement to a credit. At first sight it would seem surprising if the court could be compelled to investigate the position as to liability in respect of unrelated hereditaments. The NDR Collection Regulations 1989 do provide some support for such an idea, however, as they provide that overpayments may be credited by the billing authority to any account held by the same ratepayer.[59] This suggests that payments and liabilities in respect of different periods and hereditaments are not to be held in separate 'silos' with no interaction between them. It mirrors the finding of the House of Lords in a slightly different context; that overpayments are to be treated as

---

56 Non-Domestic Rating (Collection and Enforcement) (Local List) Regulations 1989, reg 12(2).

57 *Hardy v Sefton MBC* [2006] EWHC 1928 (Admin) at [50].

58 *Shillito v Hinchcliffe* [1922] 2 KB 236 per Shearman J at p248 relying on a line of authority going back to *Blackpool and Fleetwood Tramroad Co v Bispham with Norbreck UDC* [1910] 1 KB 592 and *London and North Western Railway Co v Bedford* (1852) 17 QB 978.

59 Non-Domestic Rating (Collection and Enforcement) (Local List) Regulations 1989, reg 9(4)(b).

having been made to a 'single indivisible fund in the hands of the [billing] authority'.[60] Cases decided under the previous provisions also support the argument that a Magistrates' Court may inquire whether a ratepayer is in truth entitled to a repayment 'in law or equity', and if so, to set that amount off against a current demand.[61] It therefore seems that a ratepayer may be able to raise an entitlement to a refund as a defence to an application for a liability order, although the point remains to be clarified in litigation. If a ratepayer is entitled to raise such a defence, the Magistrates' Court should apply the principles that would otherwise be applied by the County Court or High Court in determining whether a ratepayer is due a refund.[62]

COSTS AWARDS IN THE MAGISTRATES' COURT

Can the court make a costs order?

5.34   The powers of the Magistrates' Court to award costs are more limited than those of the ordinary civil courts. Whether or not the court has the ability to make a costs order will depend on the particular circumstances. These are considered in turn in the following paragraphs.

5.35   If the court is making a liability order, then it not only can but must include in that order 'a sum of an amount equal to the costs reasonably incurred by the [billing authority] in obtaining the order'.[63] In Wales, these costs cannot exceed a prescribed amount, which at present is £70.

---

60 *R v Tower Hamlets LBC ex p Chetnik Developments Ltd* [1988] AC 858 per Lord Bridge at p880.

61 *Stanley v Weardale Coal and Coke* (1935) 22 R&IT 279, in which a distress warrant was refused on the basis that the ratepayer had made out his disputed claim to a repayment. See also, *Wessex Electricity Co v Sherbourne UDC* (1936) 25 R&IT 136, where the ratepayer did not make out his claim to a repayment; Du Parcq J at p139 held that 'these Appellants would have had no right in law or equity to recover the sum . . . and it seems to me to follow that they have no right either in law or in equity to be credited with that sum against this further sum'. The implication is that if they did have such a right they would have been so entitled. This was separate to the question of good faith, which arises when there is no dispute that the repayment is due.

62 As to which, see Chapter 7.

63 Non-Domestic Rating (Collection and Enforcement) (Local List) Regulations 1989, reg 12(6).

5.36 If the sum subject of the liability order is paid after the order has been sought but before it has been made, the billing authority may require the court to make an order for the costs reasonably incurred in making the application.[64] These are again limited in Wales to £70.

5.37 If the court dismisses an application for a liability order, the only costs order the court can make is for the billing authority to pay the ratepayer's costs. This is unfortunate because it means that there is no penalty to a ratepayer who causes a billing authority to incur significant costs by, for example, revealing crucial information only at the last minute. The court has a discretion as to whether to make such an order and, if it does so, as to the amount. It may make 'such order . . . as it thinks just and reasonable'.[65]

5.38 If the proceedings are withdrawn by the billing authority, the only costs order the court can make is for the billing authority to pay the ratepayer's costs. Again, the court has a discretion as to whether to make such an order and, if it does so, as to the amount. It may make 'such order . . . as it thinks just and reasonable'.[66]

5.39 The court has no power to make costs orders at any other stage in the proceedings. It cannot therefore order one party to pay the other's costs in respect of an unnecessary directions hearing, for example. Such matters can only be reflected in the terms of any costs order that is made at the end of the proceedings.

Costs orders against ratepayers

5.40 As set out above, if the court is making a liability order then it *must* include the costs reasonably incurred by the billing authority in obtaining the order (or, if the sum is paid before an order is made, in making the application). This means that the only question for the court to resolve is what amount of costs have been reasonably incurred – subject to the limit of £70 in Wales. The means or ability of the ratepayer to pay those costs is not a relevant consideration, and there is no discretion to award any other amount than the sum reasonably

---

64 Non-Domestic Rating (Collection and Enforcement) (Local List) Regulations 1989, reg 12(7).

65 Magistrates' Court Act 1980, s64(1).

66 Courts Act 1971, s52(3)(b).

incurred.[67] There is no requirement that the costs incurred should be reasonable in amount or should be proportionate to the sums sought to be recovered – although if they are not this may well help to demonstrate that they were not 'reasonably incurred'.

5.41   The court must accordingly be satisfied of three facts in relation to any costs that are claimed by the billing authority:

(1) that the local authority has actually incurred those costs;
(2) that the costs in question were incurred in obtaining the liability order; and
(3) that it was reasonable for the local authority to incur them.[68]

5.42   In a contested case where the billing authority has legal representation these requirements will generally be satisfied by the provision of a costs schedule setting out the amounts incurred.

5.43   A more contentious area has been in respect of bulk lists, where what are effectively standard figures are imposed in respect of costs. The High Court has given 'general guidance' on the application of the tests, which is not binding but will be regarded as very important by the Magistrates' Court in deciding questions of costs. This guidance is to the following effect.[69] The costs must be genuinely connected with the enforcement process. As such, the only costs which will be recoverable are: (i) the costs of issuing the reminder notice; and (ii) any costs incurred after the decision to enforce by applying for a liability order. Furthermore, it may be acceptable for a billing authority to calculate a standard figure for costs by dividing the total costs incurred in the previous year by the total number of summons issued in that year, or by the number of summons predicted to be issued in the current year. In practice this is the only way that the costs of bulk lists can realistically be dealt with.

---

67 *R (Nicolson) v Tottenham Magistrates* [2015] EWHC 1252 (Admin) at [25] and [30] – a case on the identically worded council tax provisions. The High Court suggested in *Thaker v Tameside MBC* [2011] EWHC 2354 (Admin) that information as to means might be relevant in exceptional circumstances, but the relevant provision from the regulations does not appear to have been cited to it on that occasion.
68 *R (Nicolson) v Tottenham Magistrates* [2015] EWHC 1252 (Admin) at [34].
69 *R (Nicolson) v Tottenham Magistrates* [2015] EWHC 1252 (Admin) at [36], [42]–[43] and [46].

Costs orders against the billing authority

5.44 When considering a costs order against a billing authority, the court always has a discretion as to whether to make that order and as to what amount to include. It must consider what is 'just and reasonable'. Thus it has a much broader remit than when considering costs orders against ratepayers.

5.45 The relevant principles are as follows:

> What the court will think just and reasonable will depend on all the relevant facts and circumstances of the case before the court. The court may think it just and reasonable that costs should follow the event, but need not think so in all cases covered by the subsection.
>
> Where a [ratepayer] has successfully challenged before justices [a summons issued by a billing authority] acting honestly, reasonably, properly and on grounds that reasonably appeared to be sound, in exercise of its public duty, the court should consider, in addition to any other relevant fact or circumstances, both (i) the financial prejudice to the particular [ratepayer] in the particular circumstances if an order for costs is not made in his favour; and (ii) the need to encourage public authorities to make and stand by honest, reasonable and apparently sound administrative decisions made in the public interest without fear of exposure to undue financial prejudice if the decision is successfully challenged.[70]

5.46 This principle plainly will not protect a billing authority unless it has acted 'honestly, reasonably, properly and on grounds that reasonably appeared to be sound'. Where a billing authority has done that, however, then a costs award against it will be unlikely, unless the particular ratepayer in question would suffer severe financial prejudice if there were no order for costs against the billing authority.

APPEALS FROM THE MAGISTRATES' COURT

5.47 The appropriate route of appeal from a decision of the Magistrates' Court to make or refuse a liability order is an appeal by way of case

---

70 *Bradford Metropolitan District Council v Booth* [2000] EWHC (Admin) 444 as applied to the rating context in *Patel v LB of Camden* [2013] EWHC 2459 (Admin) at [15], [46].

stated to the High Court.[71] The grounds of such an appeal are that the court's decision was 'wrong in law or in excess of jurisdiction'. It is thus not an opportunity to reargue any factual disputes that were dealt with in the Magistrates' Court (unless the court has made a very egregious error in its consideration of matters of fact, or has made a finding which there was no evidence to support) nor to introduce new evidence.

5.48   The procedure for making such an appeal is governed by the Magistrates' Court Act 1980, the Magistrates' Court Rules 1981 and the Civil Procedure Rules 1998. It is frequently assumed, because the appeal is from the Magistrates' Court, that the criminal procedure rules apply, but they do not.

5.49   The applicable procedure is as follows:

5.50   Within 21 days after the day on which the decision of the Magistrates' Court was given, the party seeking to appeal must apply to the court asking it to state a case for the opinion of the High Court on the question of law or jurisdiction involved.[72] There is no particular form for making this application; it must simply 'be made in writing and signed by or on behalf of the applicant and shall identify the questions of law or jurisdiction on which the opinion of the High Court is sought'. If it seeks to ask whether there was evidence on which the decision could have been reached, it must specify the particular finding(s) of fact which it is claimed cannot be supported;[73]

5.51   Within 21 days of receipt of the application, the justices shall either issue a draft case to the parties or refuse to state a case.[74] There is a little used provision which allows the justices to demand that the applicant enter into a recognisance with or without sureties before they will state a case.[75] A recognisance is essentially a formal promise to do something, backed by an acknowledgement that a certain sum of money will be forfeit if the promise is not performed. A recognisance with sureties is similar, but involves a third party offering to pay the sum specified if the promise is not performed. The recognisance that can be required is one 'to prosecute the appeal without delay and to submit to the judgment of the High Court and pay such costs as that Court may award'.

5.52   The justices may refuse to state a case if the application is 'frivolous'.[76] In this case, the would-be Appellant will have to judicially review the

---

71 Magistrates' Court Act 1980, s111.
72 Magistrates' Court Act 1980, s111(1)–(2).
73 Magistrates' Court Rules 1981, r76.
74 Magistrates' Court Rules 1981, r77.
75 Magistrates' Court Act 1980, 114.
76 Magistrates' Court Act 1980, s111(5).

refusal to state a case. If the Magistrates' Court has already given its reasons and there appears to be an arguable ground of challenge to the decision taken, the High Court is likely to treat the claim for judicial review as a challenge to the substantive decision rather than going through the formality of requiring the justices to state a case.[77]

5.53   Where the magistrates do issue a draft case, each party has 21 days to make representations in writing on it.[78]

5.54   Within a further 21 days the justices shall state and sign the final case. The clerk may sign on behalf of the justices. It is then to be sent to the applicant (who from now on becomes known as 'the Appellant', and the other party 'the Respondent').[79]

5.55   Within 10 days of the date of the case stated, the Appellant must file his appeal (i.e. a notice of appeal stating grounds of appeal together with the case stated and any judgment given by the Magistrates' Court) with the High Court.[80] The High Court has power to extend this time limit.[81] Within four days of filing his appeal, the Appellant must serve the appeal documents on all respondents to the appeal.[82]

5.56   A respondent seeking to uphold the decision of the Magistrates' Court on grounds other than those given by the justices must file a respondent's notice within 14 days of being served with the appeal, and must serve it as soon as practicable and in any event within seven days.[83]

5.57   The High Court has extensive powers when hearing an appeal. It shall:

(a)   reverse, affirm or amend the determination in respect of which the case has been stated; or
(b)   remit the matter to the Magistrates' court . . . with the opinion of the High Court, and may make such other order in relation to the matter (including as to costs) as it thinks fit.[84]

5.58   As such the High Court can reach its own view as to liability and make or refuse a liability order accordingly. It is unlikely to do this unless the

77   *Sunworld Ltd v Hammersmith and Fulham London Borough Council* [2000] 1 WLR 2102 at 2016F–H.
78   Magistrates' Court Rules 1981, r77.
79   Magistrates' Court Rules 1981, r78.
80   Civil Procedure Rules 1998, r52.2 and CPR Practice Direction 52E para 2.2.
81   Civil Procedure Rules 1998, r52.6.
82   CPR Practice Direction 52E para 2.4.
83   Civil Procedure Rules 1998, r52.5.
84   Senior Courts Act 1981, s28A.

facts are clear. If the facts are unclear then it is more likely to remit the matter to the Magistrates' Court with its opinion. It also has the ability to deal with the costs of the Magistrates' Court proceedings, in line with the principles set out above. The costs of the appeal proceedings in the High Court will normally follow the event (i.e. the loser in the appeal will usually have to pay the winner in the appeal).

5.59 The decision of the High Court on such an appeal is final.[85] As such there is no further appeal to the Court of Appeal or Supreme Court. This is in contrast to the position in judicial review proceedings. The court will not allow judicial review proceedings which are brought purely to secure a right of appeal to the Court of Appeal, however.[86] Although it has sometimes allowed challenges to decisions of the Magistrates' Court to be brought by way of judicial review, these appear to be the exception and the normal rule is that permission for judicial review will be refused on the basis that there is an alternative remedy – i.e. appeal by way of case stated.[87] A party aggrieved by the decision of the magistrates is therefore well advised to use the case-stated procedure rather than judicial review.

5.60 As noted above, judicial review *will* be appropriate where the Magistrates' Court refuses to state a case. This leads to the bizarre situation where an Appellant has more appeal rights if the court refuses to state a case than if it does not. There is no particular logic to this situation but the relevant rules appear to be well established.

SETTING ASIDE LIABILITY ORDERS

5.61 A very large number of liability orders is made each year. It is inevitable that some mistakes will be made. It is therefore surprising that there is no provision in LGFA 1988 or the NDR Collection Regulations 1989 for liability orders made in error to be set aside.

---

85 Senior Courts Act 1981, s28A(4), s18(1)(c).
86 *Kenya Aid Programme v Sheffield City Council* [2013] EWHC 54 (Admin) at [53]–[54].
87 See *R v Barking and Dagenham LBC ex p Magon* [2004] RA 269; *R (Brighton and Hove Council) v Brighton and Hove Justices* [2004] EWHC 1800 (Admin) at [23]. In *R (Tshibangu) v Merton LBC* [2006] EWHC 2571 (Admin) at [23] the court suggested that challenges raising procedural grounds might be brought by judicial review.

5.62 If a billing authority accepts the error and wishes to make another party liable for the rates instead, it is common practice for the liability order already obtained simply to be ignored. There is no prohibition on seeking more than one liability order in respect of the same period of liability, and therefore no barrier to this course of action. The willingness of a billing authority to waive enforcement of a liability order is no reason to refuse to set it aside if it is challenged, however.[88]

5.63 There are also cases, however, where the ratepayer considers that a liability order was made in error whereas the billing authority wishes to enforce the order that has been made. To deal with such cases, the courts have accepted that the Magistrates' Court has power to set aside a liability order it has made.[89] It is an exceptional power, to be exercised cautiously and only where the ratepayer can show that all three of the following requirements are satisfied:

(1) there is a genuine and arguable dispute as to the defendant's liability for the rates in question;

(2) the order was made as a result of a substantial procedural error, defect or mishap; and

(3) the application to the justices for the order to be set aside is made promptly after the defendant learns that it has been made or has notice that an order may have been made.[90]

5.64 These three requirements are considered in turn, below.

5.65 The procedure for making such an application is simple. The application to have the liability order set aside is to be made by writing to the court, copying in the billing authority. The court will then list a hearing at which the application can be considered. If the matter is complex then it may well be that the first hearing is a preliminary hearing at which the issues can be clarified and directions made for the full hearing of the application. If there are disputes of fact, for example, as to whether certain documents were received by the ratepayer,

---

88 *R (Tull) v Camberwell Green Magistrates' Court* [2004] EWHC 2780 (Admin) at [33]–[34]. This was a council tax case but the remarks are equally relevant here.

89 This power was first accepted, or invented, in the case of *Liverpool City Council v Pleroma Distribution Ltd* [2002] EWHC 2467 (Admin).

90 Set out in *R (Brighton and Hove City Council) v Brighton and Hove Justices* [2004] EWHC 1800 (Admin) at [31] and applied in numerous cases since.

then it may well be necessary for the court to hear oral evidence and cross-examination before deciding the application.[91]

Genuine arguable dispute

5.66   It is not enough for a ratepayer to assert that he is not liable; there must be a dispute which is genuine and arguable. This has been summarised as a requirement to provide material which shows that there is a realistic prospect that the ratepayer could succeed in resisting an application for a liability order.[92] It is not necessary for the court to conclude that the ratepayer is definitely *not* liable, and the court will try to avoid being drawn too far into the merits of the ratepayer's arguments on an application to set aside a liability order.

Procedural failure

5.67   The most obvious example of a substantial procedural defect, error or mishap is where the court itself or the billing authority has made a mistake in the operation of its own procedures.

5.68   An example of an error by the court arose in the very first case to recognise the jurisdiction of the Magistrates' Court to set aside liability orders, *Liverpool City Council v Pleroma Distribution Ltd.*[93] In that case the ratepayer had written to the court seeking an adjournment of the liability order hearing, but the court staff had failed to put that letter before the court so that it could consider whether to grant an adjournment or not.

5.69   The most common error by the billing authority is failure to serve documents correctly. It is well established that if a summons has not been served correctly, there is no jurisdiction to make a liability order.[94] Failure to serve a summons in accordance with the statutory provisions[95] will therefore automatically satisfy this requirement. It would seem that failure to serve a demand notice will also satisfy this

---

91 *R (on the application of Jones) v Justices of the Peace* [2008] EWHC 2740 (Admin) at [7]. This was a council tax case but the remarks are equally relevant here.

92 *Tower Hamlets LBC v Rahman* [2012] EWHC 3428 (Admin) at [10].

93 [2013] EWHC 1921 (Admin).

94 *Chowdhury v Westminster CC* [2013] EWHC 1921 (Admin) at [28].

95 See para 5.21 on service.

requirement, as the NDR Collection Regulations 1989 provide that no payment need be made 'unless a notice *served under this Part* requires it'.[96] In those cases where it is required,[97] a reminder notice is an essential requirement for the service of a summons,[98] so again failure to serve such a notice is probably enough to satisfy this requirement.

5.70   Cases of fault by the court or billing authority provide particularly clear examples of facts which will satisfy this requirement. There is, however, no necessity for any fault on behalf of the court or the billing authority before this requirement can be satisfied.[99] In *Pleroma* the court gave the example of a ratepayer who was prevented from attending court by a traffic accident; as the Court of Appeal has observed, this is an example of a ratepayer failing to attend 'through no fault of his own'.[100] Another, perhaps more common example, would be where a summons or hearing notice has been correctly served, but simply not received by the ratepayer.[101]

5.71   At the other end of the spectrum will be cases where the failure to attend is entirely the fault of the ratepayer or his advisor. A judgment obtained in the party's absence in these circumstances will not be set aside as there has been no breach of the rules of natural justice.[102] An example from the cases is where the ratepayer's solicitor put the wrong date in their diary, and so attended court the day after the hearing.[103] It may be that some cases of non-receipt of summonses may also be attributable to the ratepayer's own fault – if, for example, he did not have any proper system for ensuring post was forwarded to him and/or attended to. In such a case the requirement will also probably not be satisfied.

96 Non-Domestic Rating (Collection and Enforcement) (Local List) Regulations 1989, reg 7(6).
97 See paras 5.18–19 on reminder notices.
98 Non-Domestic Rating (Collection and Enforcement) (Local List) Regulations 1989, reg 12(1).
99 *R (Brighton and Hove City Council) v Brighton and Hove Justices* [2004] EWHC 1800 (Admin) at [32].
100 *R (on the application of Mathialagan) v Southwark LBC* [2004] EWCA Civ 1689 at [37].
101 *R (Newham) v Stratford Magistrates' Court* [2008] EWHC 125 (Admin) at [17].
102 *Al Mehdawi* [1990] 1 AC 876.
103 *R (on the application of Mathialagan) v Southwark LBC* [2004] EWCA Civ 1689 at [37].

Prompt application

5.72   This requirement is similar in effect to the limitation periods applicable to bringing civil proceedings; it is necessary so that those limitation periods cannot be circumvented by an application to the Magistrates' Court.[104] As with limitation periods, the justification for it rests on two principles: fairness to the billing authority, which might otherwise be prejudiced by being unable to assemble evidence which would have been available had the matter been contested the first time round;[105] and the public interest in the finality of litigation.[106]

5.73   Time runs from when 'the defendant learns that it has been made or has notice that an order may have been made' and 'normally requires action within days or at most a very few weeks, not months, and certainly not as much as a year'.[107] Furthermore, 'the responsibility rests fully, fairly and squarely with the ratepayer, not simply to dispute or put in contention the justification for the liability order but to actually initiate the proceedings promptly in order to enable them to be set aside'.[108] It is therefore not enough for a ratepayer to write to the billing authority to dispute the liability order; he must actually apply to the court in order to stop time running against him and establish that he has acted promptly.

5.74   The reference above to notice that an order may have been made is specific to each order. Therefore if an order is made for one year, and the ratepayer becomes aware of it, this is not sufficient to start time running against him in respect of other orders made for other years, of which he has not had notice.[109]

HIGH COURT/COUNTY COURT PROCEEDINGS

5.75   It is possible to seek to recover unpaid rates in the County Court or High Court.[110] Such proceedings cannot be taken if a liability order

---

104   *R (Brighton and Hove City Council) v Brighton and Hove Justices* [2004] EWHC 1800 (Admin) at [33].
105   *R (Sangha) v Stratford Magistrates' Court* [2008] EWHC 2979 at [38].
106   *R (Newham) v Stratford Magistrates' Court* [2008] EWHC 125 (Admin) at [20].
107   *Brighton and Hove CC v Brighton and Hove Justices* [2004] EWHC 1800 (Admin) at [31]–[33].
108   *Sleekmade Property Company v Sheffield CC* [2015] EWHC 4193 (Admin) at [30].
109   *Sleekmade Property Company v Sheffield CC* [2015] EWHC 4193 (Admin) at [33]–[34].
110   Non-Domestic Rating (Collection and Enforcement) (Local List) Regulations 1989, reg 20.

has been made, and once they are instituted no application for a liability order may be made. The two routes are therefore alternatives.

5.76 There are no special procedural requirements for the taking of proceedings in the normal civil courts once money has fallen due pursuant to a demand notice. The proceedings will be ordinary civil proceedings brought under Part 7 of the Civil Procedure Rules 1998 ('the CPR'). The procedures to be followed will be familiar to litigation lawyers but fall outside the scope of this book.

5.77 The main advantage of proceeding by this route is that the CPR offer a much more sophisticated and flexible procedural regime than is available in the Magistrates' Court. This allows the court to manage the progress of the case more effectively and, for example, order disclosure of relevant documents by the ratepayer. The decision-makers involved may also be more suitable for complex or high-value cases, as there is no prospect of consideration by a bench of lay justices. The final advantage relates to routes of appeal. In the Magistrates' Court, no appeal may be brought beyond the High Court.[111] If proceedings are started in the High Court, however, then appeal potentially lies to the Court of Appeal and potentially to the Supreme Court.

5.78 The disadvantage is that this route will be more costly than proceeding in the Magistrates' Court and will require greater involvement from lawyers. Furthermore, the costs rules are not weighted in favour of the billing authority as they are in the Magistrates' Court. As such, it would seem only to be appropriate for a billing authority to use this procedure in the most complex and high-value cases, or those raising points of law of general public importance.[112]

FURTHER RECOVERY ACTION

5.79 In some cases, the making of a liability order may be sufficient to secure payment. In those cases where it is not, the billing authority has various options available to it. There is no limitation period in taking further enforcement proceedings once a liability order has been made.[113]

---

111 On which, see para 5.59.
112 As it was, for example, in *North Somerset DC v Honda Motor Europe Ltd and others* [2010] EWHC 1505.
113 *Bolsover DC v Ashfield Nominees Ltd* [2010] EWCA Civ 1129.

Enforcement by taking control of goods

5.80   Once a liability order has been made, the regulations allow for it to be enforced by the procedure for 'taking control of goods' in Schedule 12 of the Tribunals, Courts and Enforcement Act 2007.[114] This involves an authorised 'enforcement agent' (who would previously have been known as a 'bailiff') taking the ratepayer's goods and selling them in order to meet the debt.

Commitment to prison

5.81   A further remedy is open to the billing authority against individuals. Where the billing authority has sought to enforce payment by taking control of goods, and the enforcement agent reports that he was unable to find sufficient goods of the ratepayer to enforce payment, then the billing authority may apply for the ratepayer to be committed to prison.[115] Payment of the sum will avoid the need to go to prison or secure release from prison as the case may be, and part payment will result in the term of imprisonment being reduced in proportion to the amount paid.[116]

5.82   The Magistrates' Court when faced with such an application must conduct what is generally known as a 'means inquiry'. This involves considering both the ratepayer's means and whether the failure to pay the rates which led to the making of the liability order was due to the ratepayer's 'wilful refusal or culpable neglect'. The inquiry must be conducted in his presence.[117] There are numerous cases which deal with the principles and procedure to be applied to such an application, many of which were decided in the context of the community charge (or 'poll tax' as it was generally known). The legal principles would appear to be the same, however, so those cases are of equal relevance here.

114 Non-Domestic Rating (Collection and Enforcement) (Local List) Regulations 1989, reg 14.
115 Non-Domestic Rating (Collection and Enforcement) (Local List) Regulations 1989, reg 16.
116 Non-Domestic Rating (Collection and Enforcement) (Local List) Regulations 1989, reg 16(7).
117 Non-Domestic Rating (Collection and Enforcement) (Local List) Regulations 1989, reg 16.

5.83 The inquiry that the court is required to make must focus on the period between the demand notice and the liability order; that is, the period within which it must be shown that failure to pay was caused by 'wilful refusal or culpable neglect'. As such, if the Magistrates' Court fails to consider this period and simply looks at the ratepayer's conduct and means since the making of the liability order, its decision will be quashed.[118] Because the possible result of the inquiry is imprisonment, the higher courts expect that it will be thorough. It must be a 'full and proper' inquiry,[119] and the justices must 'approach the matter with considerable care and caution because the order is one which has serious implications for the [ratepayer], for the rating authority and for the public'.[120] It follows that if there has been no meaningful inquiry at all, the decision will be quashed.[121]

5.84 The court will not generally have to consider whether there are alternative means of securing payment than commitment to prison, but in exceptional cases should do so: for example, where the debtor has a low income but large capital assets which if sold could pay the debt.[122]

5.85 The procedure of a means inquiry is relatively formal. A ratepayer is entitled to have the assistance of a friend in dealing with a means inquiry, and to cross-examine the billing authority's officer. This cross-examination can relate to the preconditions of making the order, i.e. what amount is due and whether attempts have been made to collect it by taking control of goods,[123] or to any evidence as to means that comes from a source other than the ratepayer.[124]

---

118 *R v Manchester City Magistrates' Court ex p Davies* [1989] QB 631. Note that there is no longer any right to claim damages for false imprisonment unless, as well as acting without jurisdiction, the justices have also acted 'in bad faith': Courts Act 2003, s32(1). See also *R v Newcastle upon Tyne ex p Devine* [1998] RA 97 at p101–102.

119 *R v Richmond Justices ex p Atkins* [1983] RVR 148, in which the justices wrongly assumed that the debtor could secure a further advance from his bank.

120 *R v Birmingham Magistrates' Court ex p Mansell* [1988] RVR 112 per Woolf LJ at p114.

121 *R v Birmingham Justices ex p Turner* (1971) 17 RRC 12 and *R v Liverpool City Justices ex p Lanckriet* [1977] RA 85 are both examples of this. It seems unlikely that a Magistrates' Court would ever act in this fashion today.

122 *R v Birmingham Magistrates' Court ex p Mansell* [1988] RVR 112.

123 *R v Highbury Corner Magistrates' Court ex p Watkins* [1992] RA 300 at p304–305.

124 *R v Wolverhampton Magistrates' Court ex p Mould* [1992] RA 309 at p319.

5.86   A ratepayer can adduce evidence and argument about liability as part of his case at such an inquiry.[125] It has been suggested that such evidence would be inadmissible because the making of the liability order was 'water under the bridge'; those views appear to have been expressed without knowledge of an earlier decision allowing reference to be made to liability, and furthermore are not a binding authority.[126] If the issue of liability has been resolved at a contested hearing it seems very unlikely that a Magistrates' Court would be prepared to re-open the dispute at the committal stage. A genuine dispute as to liability might be more relevant where the liability order has been made in the ratepayer's absence; it would tend to explain why there has been no payment and therefore to count against a finding that the lack of payment was due to 'wilful refusal or culpable neglect'.

5.87   If satisfied of the ratepayer's 'wilful refusal or culpable neglect', then the court has a discretion as to what order to make. It can commit the ratepayer to prison immediately, or it can fix a term of imprisonment but postpone the commencement of that term on conditions.[127] If it does neither of these things it can remit all or part of the sum due.[128] If no order is made then the billing authority can renew the application later if circumstances change.[129]

5.88   As stated above, the court has a discretion. This means that it is not obliged to make an order even if satisfied of 'wilful refusal or culpable neglect'. An order is 'only to be made if payment can be made and there is no other way of inducing the ratepayer to do so'.[130] If the magistrates fail to appreciate that they have a discretion and therefore fail to exercise it, any order is likely to be quashed.[131]

---

125 *R v Ealing Magistrates' Court ex p Coatsworth* [1980] RA 97 is authority for this point.

126 *R v Highbury Corner Magistrates' Court ex p Watkins* [1992] RA 300 at p305.

127 Non-Domestic Rating (Collection and Enforcement) (Local List) Regulations 1989, reg 16(3).

128 Non-Domestic Rating (Collection and Enforcement) (Local List) Regulations 1989, reg 17(2).

129 Non-Domestic Rating (Collection and Enforcement) (Local List) Regulations 1989, reg 17(3).

130 *R v Oundle and Thrapston Justices ex p East Northamptonshire DC* [1980] RA 232 at p236.

131 As happened in *Stevenson v Southwark LBC* [1993] RA 113.

In exercising the discretion, they should consider whether the rate-payer 'had the ability to pay and [is] deliberately and unjustifiably withholding payment'.[132]

5.89 It is relevant to the court's discretion that imprisonment is 'plainly intended to be used as a weapon to extract payment rather than to punish'.[133] It is coercive not punitive. However, the court is plainly entitled to use imprisonment as a punitive measure as a last resort, because 'if the coercive regime is seen to have no teeth, then it will soon lose any coercive force'.[134] The predominant purpose, however, is coercive not punitive.[135]

5.90 The period of imprisonment that can be imposed is limited to three months.[136] This remains the case even if the application relates to multiple properties or periods of non-payment; the court cannot impose consecutive periods of imprisonment in respect of different sums so as to extend the overall period beyond three months.[137] The principle of proportionality is relevant when fixing a term of imprisonment: 'the more serious the case, whether in terms of the amount outstanding or in terms of the degree of culpability or blame to be attached to the ratepayer for his non-payment, the closer will any period imposed approach the maximum'.[138] It follows that the period of imprisonment will usually be less, all other things being equal, in a case of culpable neglect than in a case of wilful refusal.

5.91 It is common for the term of imprisonment to be suspended on condition that the ratepayer pay the amount by a certain date or in instalments. This obviously improves the coercive effect of the order and should normally be the first choice for magistrates faced with an application for committal to prison. If the term is suspended on condition of regular payments, those payments should not extend

---

132 *Stevenson v Southwark LBC* [1993] RA 113 at p122.
133 *R v Wolverhampton Magistrates' Court ex p Mould* [1992] RA 309 at p330. Applied in *R v Middleton Justices ex p Tilley* [1995] RVR 101.
134 *R v Cannock Justices ex p Ireland* [1996] RA 463 at p468.
135 *Smith v Braintree DC* [1990] 2 AC 215 at p230.
136 Non-Domestic Rating (Collection and Enforcement) (Local List) Regulations 1989, reg 16(7).
137 *R v Bexley Justices ex p Henry* (1971) 17 RRC 15.
138 *R v Highbury Corner Magistrates' Court ex p Uchendu* [1994] RA 51 at p54.

over an unreasonable period and 'certainly not a period in excess of three years'.[139]

5.92   If a suspended term of imprisonment is ordered and the conditions are breached, committal to prison for the specified term may follow. Imprisonment is not automatic and is not to be ordered by the court of its own motion. The billing authority must apply to have the debtor committed. The court must inquire as to whether new failure to pay is due to wilful refusal or culpable neglect.[140] It is the new failure that is relevant, because the court has already been satisfied as to the failure to pay which led to the making of the liability order. The inquiry should therefore be limited to considering whether the ratepayer had the ability to pay the instalments but failed to do so; if his circumstances have changed such that he cannot pay, it would not be just to commit him to prison.[141]

5.93   Where a ratepayer breaches the conditions on which imprisonment was suspended, and the billing authority applies to have him committed to prison, natural justice requires that he be given an opportunity to be heard in response to the application.[142] An *opportunity* is all that is required, however. Unlike the initial means inquiry, there is no absolute requirement that the ratepayer be present. If he is given notice of the hearing but fails to attend, the court is entitled to commit him to prison without hearing from him.[143]

5.94   Commitment to prison has declined in popularity as a remedy for billing authorities. This may be because the costs the billing authority can recover in respect of an application are limited by the regulations.[144] Furthermore, committal to prison and the more flexible remedy of bankruptcy are mutually exclusive remedies for the billing authority, and both cannot be pursued at once.[145] If another party has made the

---

139   *R v Newcastle upon Tyne ex p Devine* [1998] RA 97 at p103.

140   *R v Poole Justices ex p Fleet* [1983] 1 WLR 974.

141   *R v Felixstowe etc. Magistrates' Courts ex p Herridge* [1993] RA 83 at p93.

142   *R v Poole Justices ex p Fleet* [1983] 1 WLR 974.

143   *R v Northamptonshire Magistrates' Court ex p Newell* [1992] RA 283.

144   Non-Domestic Rating (Collection and Enforcement) (Local List) Regulations 1989, reg 16(7) and sch 4. The current limit is £305 for making an application for a warrant of commitment and £145 for making an application for a warrant of arrest.

145   Non-Domestic Rating (Collection and Enforcement) (Local List) Regulations 1989, reg 19(2).

ratepayer bankrupt, then committal proceedings can and usually will be stayed in order to protect his estate for his creditors.[146]

Insolvency

5.95   A liability to pay non-domestic rates, whether or not it is the subject of a liability order, is regarded as a debt for the purposes of insolvency proceedings.[147] As such, it can provide the basis for a billing authority to:

(1)   petition for the winding up of a company;
(2)   seek an order putting a company in administration; or
(3)   petition for an individual to be made bankrupt.

5.96   The rules governing these regimes are outside the scope of this book.

CENTRAL LIST HEREDITAMENTS

5.97   Recovery of rates in respect of central list hereditaments is governed by the Non-Domestic Rating (Collection and Enforcement) (Central Lists) Regulations 1989.

5.98   The basic structure of those regulations in terms of billing is similar to the local list regulations. There is a requirement for demand notices to be served to crystallise liability to pay.[148] These notices must be served 'as soon as practicable' after the relevant date,[149] and there are similar provisions on the service of notices.[150] Payments will be by instalments unless otherwise agreed and where there is a failure to pay instalments

---

146   *Smith v Braintree DC* [1990] 2 AC 215.
147   Non-Domestic Rating (Collection and Enforcement) (Central Lists) Regulations 1989, reg 18(1)–(2) for liability orders, and see *Bolsover DC v Ashfield Nominees Ltd* [2010] EWCA Civ 1129 at [34] in respect of sums not yet the subject of liability orders.
148   Non-Domestic Rating (Collection and Enforcement) (Central Lists) Regulations 1989, reg 4.
149   Non-Domestic Rating (Collection and Enforcement) (Central Lists) Regulations 1989, reg 5.
150   Non-Domestic Rating (Collection and Enforcement) (Central Lists) Regulations 1989, reg 3.

the whole amount becomes payable.[151] There is provision for making final adjustments where the amount demanded is found to be different from what is due.[152]

5.99　The main difference is that liabilities in respect of central list hereditaments cannot be enforced by way of liability order in the Magistrates' Court. Instead, recovery must be in the County Court or High Court. Once again, it is not possible to challenge the list in such proceedings.[153]

151 Non-Domestic Rating (Collection and Enforcement) (Central Lists) Regulations 1989, reg 6–8.
152 Non-Domestic Rating (Collection and Enforcement) (Central Lists) Regulations 1989, reg 9.
153 Non-Domestic Rating (Collection and Enforcement) (Central Lists) Regulations 1989, reg 10.

6

Completion notices

6.1 The effect of the provisions considered in previous chapters is that rates can only become payable once a hereditament is entered in the list. In the case of new or altered buildings which are occupied straightaway, it is fairly obvious when they should be entered in the list. However, the fact that rates are payable on empty buildings has given rise to issues in relation to buildings which are completed, or partially completed, and then left empty. At what point should the new or altered building be entered in the list and become liable to empty rates? These are the issues which the regime for serving completion notices is designed to address.

6.2 This chapter first considers the normal principles of rating law as to when a building is complete and due to be entered in the list, before dealing with the scope and effect of completion notices.

NORMAL PRINCIPLES: WHEN SHOULD A NEW BUILDING BE ENTERED IN THE LIST?

6.3 There is now no requirement for a completion notice to be served before a new building can be entered into the list. This contrasts with the position under the GRA 1967, when it was held that a completion notice was essential.[1] Instead, the Valuation Officer can simply add the new building to the list under his power to maintain the list,[2] or in response to a proposal to alter the list.

1 *Watford BC v Parcourt Properties* [1971] RA 97 at p110.
2 LGFA 1988, s41(1).

6.4   This can be done as soon as the building is a hereditament. The Upper Tribunal has summarised the test for whether a new building is a hereditament as follows:

> A building is only a hereditament if it is ready for occupation, and whether it is ready for occupation is to be assessed in the light of the purpose for which it is designed to be occupied. If the building lacks features which will have to be provided before it can be occupied for that purpose and when provided will form part of the occupied hereditament and form the basis of its valuation it does not constitute a hereditament and so does not fall to be shown in the rating list. There is in consequence no scope for including in the list a building which is nearly, even very nearly, ready for occupation unless the completion notice procedure has been followed.[3]

6.5   The application of this test will be a matter of fact and degree. Some guidance is provided by the cases in that the following buildings have been found not to meet the test (or earlier iterations of it):

(1)   large offices with no partitioning;[4]
(2)   workshop unit with no electrical wiring or independent power supply;[5]
(3)   offices without small power facilities (i.e. a ring main and power points), tea points (i.e. small kitchens) and partitioning;[6]
(4)   warehouses lacking small power distribution, lighting and, in one case, a gas connection necessary to supply hot water.[7]

6.6   It is apparent that this test is a stringent one. Therefore the Upper Tribunal has repeatedly stressed the importance of using the completion notice procedure, which offers what is in theory a more reliable means of securing the addition of new buildings to the list.

---

3 *Porter (VO) v Trustees of Gladman SIPPS* [2011] UKUT 204 (LC) at [66].
4 *Watford BC v Parcourt Properties* [1971] RA 97.
5 *Spears Brothers v Rushmoor BC* [2006] RA 86.
6 *Porter (VO) v Trustees of Gladman SIPPS* [2011] UKUT 204 (LC).
7 *Aviva v Whitby (VO)* [2013] UKUT 430 (LC).

COMPLETION NOTICES

Scope of completion notice regime: 'new buildings'

6.7 A completion notice can only be served in respect of a 'new building'.[8] It will often be fairly obvious whether a particular building is a 'new building' in the sense that it has just been constructed. However, the definition of 'new building' is extended as follows:

(a) 'building' includes part of a building; and

(b) references to a new building include references to a building produced by the structural alteration of an existing building where the existing building is comprised in a hereditament which, by virtue of the alteration, becomes, or becomes part of, a different hereditament or different hereditaments.[9]

6.8 Sub-section (a) provides that 'building' includes a part of a building, so it would appear that a new extension or addition to a building would itself constitute a 'new building'.

6.9 Sub-section (b) includes a twofold test: first, there must be a 'structural alteration' of an existing building; second, this must produce a different hereditament(s). The concept of a 'structural' alteration has been criticised by the Court of Appeal in another rating context as being unsatisfactory,[10] but there are no authorities dealing with the term in this provision. A measure of common sense will have to be used in testing whether a particular alteration of a building is 'structural' or not.

6.10 There is also no guidance on the second limb of the test; namely, whether a different hereditament(s) is created. Where one existing building is structurally altered so as to form two new and non-interconnecting units, then it may be obvious that the test is passed. Otherwise, however, it is unclear to what extent the new hereditament needs to be different. It seems unlikely that a change of description and/or rateable value would be enough; probably the new hereditament would have to occupy a different area to the old in order for the building to fall within

---

8 LGFA 1988, s46A(1) and sch 4A para 1(1) and (2).

9 LGFA 1988, s46A(6).

10 *Williams (VO) v Scottish & Newcastle Retail Ltd* [2001] EWCA Civ 185 at [75].

the scope of this provision. There will inevitably be no occupation to guide the definition of the new hereditament, so there is further scope for controversy here.

When to serve a completion notice

6.11   There are two situations in which a billing authority may serve a completion notice.

6.12   The first is where it comes to the notice of the billing authority that a new building[11] in its area has already been completed.[12] The test of whether a building is complete is the same as the test of whether it is a hereditament, as set out above.[13] In this situation, the billing authority 'may' serve a completion notice. Presumably the decision is a discretionary one because the billing authority might equally leave bringing the hereditament into the list to the Valuation Officer.

6.13   The second situation is where it comes to the notice of the billing authority that the work remaining to be done on a new building in its area is such that the building can reasonably be expected to be completed within three months.[14] In this case, the billing authority 'shall' serve a notice 'as soon as reasonably practicable' unless the Valuation Officer directs otherwise. Presumably the service of a notice is mandatory in this case because the Valuation Officer has no powers to add such a building to the list unless a completion notice is served. It would appear that a consideration of whether work can reasonably be expected to be completed includes a consideration of whether a reasonable owner would in fact carry out the work.[15]

6.14   In general, then, a completion notice cannot be served unless the new building can reasonably be expected to be complete within three months from the date of service of the notice. This position is subject to modification in the case of developments which are

---

11 See paras 6.7–10 for the definition of this term.

12 LGFA 1988, sch 4A para 1(2).

13 See paras 6.3–6. The test is derived from various authorities on the completion notice provision in GRA 1967; namely, *Watford BC v Parcourt Properties* [1971] RA 97 at p105, *Ravenseft v Newham LBC* [1976] QB 464 at p474 and *Post Office v Nottingham* [1976] 1 WLR 624 at p635–636.

14 LGFA 1988, sch 4A para 1(1).

15 See *Spears Brothers v Rushmoor BC* [2006] RA 86 at [17], where the motivations of the owner are assumed to be relevant; it is not apparent whether the point was argued.

generally completed in two phases. The modification in such cases is as follows:

9—

(1) This paragraph applies in the case of a building to which work remains to be done which is customarily done to a building of the type in question after the building has been substantially completed.

(2) It shall be assumed for the purposes of this Schedule that the building has been or can reasonably be expected to be completed at the end of such period beginning with the date of its completion apart from the work as is reasonably required for carrying out the work.[16]

6.15   The House of Lords explained in *London Merchant Securities Plc v Islington*[17] that the purpose of this provision is to allow the completion notice procedure to apply in the case of speculative developments. These will be partially completed and then left in that state until an occupier has been identified, at which point the development will be fitted out to his requirements. The phrase 'work . . . customarily done to a building of the type in question after the building has been substantially completed' refers to the 'second phase' fit-out of such developments. The time 'reasonably required' to do this work must be calculated without taking account of the need to find a tenant, although 'incidental' matters such as deliveries may perhaps be included.[18]

6.16   Where paragraph 9 applies it has two consequences. The first consequence is that the completion date is fixed by reference to the date on which the building was completed apart from the fitting-out works.[19] Time will start to run against the owner from that date, not from the date the completion notice is served. In other words, if there are five months of fitting-out works still to do at the date of the notice, it will be valid as long as the building was otherwise complete two or more months before the date of the notice.

16 LGFA 1988, sch 4A para 9.
17 [1987] RA 89 at p101–103.
18 *JLG Investments v Sandwell DC* (1977) 245 EG 137 at p104.
19 *Graylaw Investments v Ipswich BC* [1979] JPL 767.

6.17 The second consequence of paragraph 9 would appear to be that no completion notice can be served until the building is complete apart from the customary works. This is because there can be no overlap between the fitting-out works and the other works.[20]

Determination of a completion day

6.18 The completion day proposed in the notice will depend on the circumstances in which the notice has been served:

(1) Where the notice is served because the building appears to be complete already, the completion day will be the date of the notice;[21]

(2) Where the notice is served because it appears that the work remaining to be done on the building can reasonably be expected to be completed within three months, the completion day will be a day by which the building can reasonably be expected to be completed;[22]

(3) Where paragraph 9 applies,[23] there is a statutory assumption about when the building 'has been or can reasonably be expected to be completed'. The completion day in the notice will therefore either be that date (as long as it is within three months of the service of the notice), or, if that date falls before the date of the notice, will be the date of the notice. This is irrespective of the actual state of completion of the building when the notice is served, or when it might actually be expected to be completed by.

6.19 The completion day is not finally determined, however, until one of the following four events occurs:

(1) The person on whom the notice is served agrees a completion day with the billing authority in writing.[24] There would seem to be no reason why such a day could not be more than three months

---

20 *London Merchant Securities Plc v Islington* [1987] RA 89 at p101–103.
21 LGFA 1988, sch 4A para 2(3).
22 LGFA 1988, sch 4A para 2(2).
23 See para 6.14–16.
24 LGFA 1988, sch 4A para 3(1).

from the date of the notice.[25] There may be advantages for the authority and the owner in agreeing such a date as a means of compromising what would otherwise result in a contested appeal against the notice. No prejudice is caused to the owner because the later date can only be imposed with his agreement.

(2) The tribunal on appeal has determined a completion day.[26] In this case, it would seem odd if the completion day could be more than three months from the service of the notice. If so, this would seem to offer the billing authority a way to extend its power to serve completion notices beyond that provided for in LGFA 1988. On balance therefore it seems that any day determined by the tribunal on appeal must be within three months of the date of service of the notice. This understanding of the provisions appears to be shared by the tribunals.[27]

(3) An appeal against the notice is dismissed or withdrawn.[28] In this case the completion day will be the day proposed in the notice.

(4) The time for an appeal expires and no appeal has been brought.[29] In this case also the completion day will be the day proposed in the notice. This is subject, however, to the tribunal's power to allow an appeal against a completion notice out of time.[30]

Effect of a completion notice

6.20 The effect of a completion notice is threefold.

6.21 In terms of valuation and the maintenance of the list, the hereditament is deemed to be completed by the 'relevant day' if it is not already

---

25 Such an agreement did not attract any comment from the court or the parties in *Camden LBC v Post Office* [1977] 1 WLR 892 (nor in the first instance judgment at (1977) 242 EG 239).

26 LGFA 1988, sch 4A para 4(2).

27 See *Spears Brothers v Rushmoor BC* [2006] RA 86 at [19], where the Upper Tribunal quashed a notice having determined that the completion day was more than three months from the date of service of the notice. The point does not seem to have been argued, however, and it is notable that the Court of Appeal appeared to see no problem with a completion date some six months from the date of the notice in *Graylaw Investments v Ipswich BC* [1979] JPL 767.

28 LGFA 1988, sch 4A para 5.

29 LGFA 1988, sch 4A para 5.

30 Valuation Tribunal for England (Council Tax and Rating Appeals) (Procedure) Regulations 2009, reg 6(3)(a).

complete by that day.[31] The relevant day is the day determined as the completion day, under the provisions dealt with above. Somewhat confusingly, however, if an appeal is brought against the notice, then the relevant day is the day stated in the notice.[32] This means that the hereditament will be entered in the list and will become liable to empty rates pending the determination of any appeal.[33] If, however, a different date is later substituted, then the list will be altered to reflect the new completion date.[34]

6.22   The second effect relates to liability to empty rates. The building is deemed to have become unoccupied for the purpose of liability to empty rates from the completion day as finally determined.[35] The wording of the provision makes it clear that if the building is subsequently entered in the list as multiple hereditaments, each of them will be deemed to have become unoccupied as from that date.[36]

6.23   The third effect only applies when the new building has been created by the structural alteration of an existing hereditament. That hereditament is deemed to cease to exist from the completion day.[37] This means that it is deemed to continue to exist until that date;[38] although it may well have only a nominal valuation whilst it is being structurally altered.[39]

Validity and service of completion notices

6.24   The procedural requirements for the contents of a completion notice are very limited. In order to be valid, it must simply comply with paragraph 2(1) of schedule 4A:

> A completion notice shall specify the building to which it relates and state the day which the authority proposes as the completion day in relation to the building.

31 LGFA 1988, s46A(2).
32 LGFA 1988, s46A(3).
33 LGFA 1988, sch 4A, para 6.
34 Non-Domestic Rating (Alteration of Lists and Appeals) (England) Regulations 2009, reg 14(3)–(4).
35 LGFA 1988, s46A(4).
36 This sensible construction was also put on the apparently more restrictive words of the GRA 1967 by the Court of Appeal in *Camden LBC v Post Office* [1977] 1 WLR 892.
37 LGFA 1988, s46A(5).
38 *Easiwork Homes v Redbridge LBC* [1970] 2 QB 406 at p416E.
39 *Ravenseft v Newham LBC* [1976] 1 QB 464 at p479B.

6.25 There is no further requirement to name the owner of the building in the notice, although it is preferable to do so.[40] The notice must do no more than 'fairly convey to the recipient what is the subject matter of the notice'.[41] Therefore a notice may in some circumstances be effective even if it does not give the correct name for the newly erected building, as long as it is in fact clear which building is being referred to.

6.26 The laxity of these procedural requirements can lead to surprising results. In *Prudential Assurance Co Ltd v VO*,[42] an agreement on the appropriate completion day reached pursuant to an earlier notice (since recognised to be invalid) was treated as a completion notice in its own right. There was no need for it to claim to be a notice or to state the recipient's appeal rights in order to have effect as a completion notice.

6.27 Before a valid completion notice can take effect, it must be served on the owner of the building to which it relates.[43] The requirements which must be complied with in order to serve a notice have proved to be contentious. However, the Upper Tribunal has clarified that the question of whether a notice has been effectively served is in essence a simple one. It relied on the following definition of 'service':

> 'Serve' is an ordinary English word connoting the delivery of a document to a particular person.[44]

6.28 As such it found that:

> Whether that mode of service was effective or ineffective did not depend on the content of the document itself, but on whether it was delivered in such a way as to come into the hands of the intended recipient.[45]

---

40 *Westminster v UKI (Kingsway) Ltd* [2015] UKUT 301 LC at [32], [36].
41 *Henderson v Liverpool MDC* [1980] RA 238.
42 [2011] RA 490.
43 LGFA 1988, sch 4A para 1(1).
44 *Tadema Holdings Ltd v Ferguson* (2000) 32 HLR 866 as cited in *Westminster v UKI (Kingsway) Ltd* [2015] UKUT 301 LC at [38].
45 *Westminster v UKI (Kingsway) Ltd* [2015] UKUT 301 LC at [41].

6.29　In other words, if a notice is in fact delivered to the intended recipient, then it will have been served, however it made its way there. Electronic service is acceptable for this reason.[46]

6.30　The LGFA 1988 does, however, provide guaranteed methods of service. If service is effected in one of these ways, then the notice will be effective whether or not the owner actually receives it. In other words, the 'risk of non-receipt' is transferred from the billing authority to the owner.[47] They are as follows:

> Without prejudice to any other mode of service, a completion notice may be served on a person—
>
> (a)　by sending it in a prepaid registered letter, or by the recorded delivery service, addressed to that person at his usual or last known place of abode or, in a case where an address for service has been given by that person, at that address;
>
> (b)　in the case of an incorporated company or body, by delivering it to the secretary or clerk of the company or body at their registered or principal office or sending it in a prepaid registered letter or by the recorded delivery service addressed to the secretary or clerk of the company or body at that office; or
>
> (c)　where the name or address of that person cannot be ascertained after reasonable inquiry, by addressing it to him by the description of 'owner' of the building (describing it) to which the notice relates and by affixing it to some conspicuous part of the building.[48]

6.31　It is important to be clear that the third method of service – namely, leaving the notice at the building addressed to the 'owner' – is *only* acceptable if the name or address of the owner 'cannot be ascertained after reasonable inquiry'. Given the relative ease with which landowners can now be traced through the Land Registry and Companies House, coupled if necessary with the billing authority's power to make statutory requests for information,[49] it would seem that this requirement will rarely be satisfied.

46　*Westminster v UKI (Kingsway) Ltd* [2015] UKUT 301 LC at [49].

47　This is how it was put in *Westminster v UKI (Kingsway) Ltd* [2015] UKUT 301 LC at [35].

48　LGFA 1988, sch 4A para 8.

49　Under s16 of the Local Government (Miscellaneous Provisions) Act 1976.

Appeal against completion notice

6.32 LGFA 1988 provides a right of appeal for those served with a completion notice. The recipient of the notice may appeal to a Valuation Tribunal.[50] The ground of appeal is simply that 'the building to which the notice relates has not been or, as the case may be, cannot reasonably be expected to be completed by the day stated in the notice'.

6.33 The time limit for such an appeal is 28 days. The start date for this period is slightly different in England and Wales. In England, time runs from the date on which the appellant received the notice.[51] In Wales, time runs from 'the service of the notice'.[52] The two dates may not be the same, because the provisions for deemed service mean that a notice can be validly served even if not in fact received.[53] The tribunal in England can however extend this time limit under its general power to regulate its own procedure.[54] In Wales, the power is more limited, and an extension may only be granted if the President of the Valuation Tribunal for Wales is satisfied that 'the failure to initiate the appeal . . . has arisen by reason of circumstances beyond [the appellant's] control'.[55] If a notice has not in fact been received until after deemed service has been effected then an application for an extension of time would seem likely to be successful.

6.34 The Valuation Tribunal's powers on an appeal against a completion notice are limited to amending the completion day.[56] It may or may not have jurisdiction to consider whether the notice is valid.[57] Even if it does have jurisdiction to do so, however, it cannot order the deletion of the hereditament from the list on the basis that the notice is invalid. As such, anyone seeking to challenge not just the completion day in the notice but the validity of the notice should also make a proposal for the deletion of the hereditament from the list.[58]

---

50 LGFA 1988, sch 4A para 4(1).
51 The Non-Domestic Rating (Alteration of Lists and Appeals) (England) Regulations 2009, reg 19(1).
52 The Non-Domestic Rating (Alteration of Lists and Appeals) (Wales) Regulations 2005, reg 19(1).
53 See para 6.30.
54 Valuation Tribunal for England (Council Tax and Rating Appeals) (Procedure) Regulations 2009, reg 6(3)(a).
55 The Valuation Tribunal for Wales Regulations 2010, reg 29(5).
56 *R (Reeves (VO)) v VTE* [2015] EWHC 973 (Admin) at [58].
57 *R (Reeves (VO)) v VTE* [2015] EWHC 973 (Admin) at [22]–[26].
58 *R (Reeves (VO)) v VTE* [2015] EWHC 973 (Admin) at [60].

Withdrawal of completion notice

6.35 There is a specific power for a billing authority to withdraw a completion notice by serving a subsequent notice.[59] This power to withdraw a notice ceases once the completion day has been determined[60] (i.e. through a decision on an appeal or by agreement).[61] Where an appeal has been brought against a notice, but has not yet concluded, the power can still be exercised with the consent of the building's owner.[62]

6.36 It is unclear whether a billing authority has a general power to withdraw a notice it has served. The existence of the provisions giving a specific and limited power to withdraw a notice would suggest that it does not. This may lead to difficulty for a billing authority if an owner who has made an appeal refuses his consent to the withdrawal of the first notice. In these circumstances it would presumably be necessary to ask the Valuation Tribunal to determine the appeal administratively on the basis that it was no longer contested.

6.37 A completion notice is also deemed to have been withdrawn if a completion day has been agreed.

---

59 LGFA 1988, sch 4A para 1(3).
60 LGFA 1988, sch 4A para 1(5).
61 See para 6.19.
62 LGFA 1988, sch 4A para 1(4).

7

Rate refunds

7.1   Unsurprisingly, the legislation contains specific provision governing the situation where there has been, for whatever reason, an overpayment of rates. These provisions may not provide a sufficient remedy for all ratepayers who have overpaid, however. It is therefore also necessary to consider in this chapter two other causes of action that do not depend on any specific provision in the regulations: namely, a claim for restitution of unlawful taxes; and a claim for restitution of money paid under a mistake.

STATUTORY ENTITLEMENT TO A REFUND

7.2   Legislative provisions on refunds are to be found in the relevant collection and enforcement regulations for central[1] and local[2] lists. The regime for central lists is in substance the same as that for local lists, which is considered below.

The entitlement

7.3   Regulation 9, which provides the entitlement to a refund, applies where three criteria are met:

(a) a billing authority has required payment from a ratepayer in respect of a hereditament for a chargeable year;

1 Non-Domestic Rating (Collection and Enforcement) (Central List) Regulations 1989, reg 9.
2 Non-Domestic Rating (Collection and Enforcement) (Local List) Regulations 1989, reg 9.

(b)   the payment required is 'found to be in excess of or less than the amount payable' in relation to the hereditament for the year; and

(c)   provision for adjusting the amounts required under the notice is not made by any other provision.

7.4   The first of these criteria is self-explanatory: there must have been an original demand notice. If the overpayment has been made otherwise than in response to a demand notice, therefore, this provision does not help the ratepayer.

7.5   The second criterion requires that the payment required is 'found to be' inaccurate. In general terms the meaning of this is obvious. The wording is wide enough to cover both a mistake in the bill when it was issued and a later change to the list which renders a bill inaccurate. The 'amount payable' is a technical term for the ratepayer's liability determined in accordance with LGFA 1988.[3] As such, it will be apparent that the regulation requires a pure exercise of calculation – what is the liability and does the payment required by the bill exceed (or understate) that? There is no discretion to dis-apply the regulation even if the excess is the result of, say, a clear error in the list.

7.6   The regulation does not specify that the payment required must be 'found to be' inaccurate by anyone in particular; on balance therefore it appears that there is no requirement for the billing authority itself to acknowledge the inaccuracy before this regulation can apply (although in most cases it is likely to be the billing authority which first spots the inaccuracy).[4] The standpoint would thus appear to be that of an objective observer aware of all the relevant facts. As such, the regulation is likely to be held to apply as soon as the inaccuracy arises, even if all parties are unaware of the error.

7.7   Finally, the reference in the third criterion to other provisions relates primarily back to provisions in the regulations for dealing with payment by instalments.[5]

---

3   Non-Domestic Rating (Collection and Enforcement) (Local List) Regulations 1989, reg 3(1); Non-Domestic Rating (Collection and Enforcement) (Central List) Regulations 1989, reg 2(1).

4   By contrast, the provisions of the General Rate Act 1967 required by s9 that the inaccurate liability must be 'shown to the satisfaction of a rating authority' before a refund could be made.

5   Non-Domestic Rating (Collection and Enforcement) (Local List) Regulations 1989, reg 7; Non-Domestic Rating (Collection and Enforcement) (Central List) Regulations 1989, reg 7.

7.8   If all three criteria are satisfied, then the regulation applies. This leads to two consequences, which would appear from the drafting of the regulations to be separate and distinct. The first is that the billing authority is required to serve an updated demand notice. If there has been an underpayment, it is empowered to rely on the corrected demand notice to secure payment of the difference.[6] The second is that, if there has been an overpayment, the overpayment becomes repayable to the ratepayer. Unlike the recovery of any underpayment, this is not made dependent on the issuing of a new demand notice. If the ratepayer requires it, the overpayment 'shall' be repaid. Otherwise, the billing authority can either repay the amount or credit it against any subsequent liability of the ratepayer.[7]

7.9   Where the billing authority makes a repayment in error, there is no provision under the regulations for the money to be recovered from the ratepayer.[8] Any claim to recover the mistaken repayment from the ratepayer would have to be made in restitution. The ratepayer will be able to advance the defence of 'change of position' if the money is no longer available to repay to the billing authority.[9]

Proceedings to secure repayment

7.10   If the billing authority fails in its obligation to refund the overpayment, the ratepayer may take proceedings in a 'court of competent jurisdiction' (i.e. the High Court or County Court) to recover it.[10] There is no defence to such an action other than to argue either that regulation 9 does not apply (because one of the three criteria is not met) or that there has been no overpayment.

---

6 Non-Domestic Rating (Collection and Enforcement) (Local List) Regulations 1989, reg 9(2)–(3); Non-Domestic Rating (Collection and Enforcement) (Central List) Regulations 1989, reg 9(2)–(3).

7 Non-Domestic Rating (Collection and Enforcement) (Local List) Regulations 1989, reg 9(4); Non-Domestic Rating (Collection and Enforcement) (Central List) Regulations 1989, reg 9(4).

8 *AEM (Avon) Ltd v Bristol* CC [1998] RA 89.

9 See para 7.30 onwards for a discussion of the legal principles relevant to restitutionary claims, although note that the claim would in this case have to be made by the billing authority not the ratepayer.

10 Non-Domestic Rating (Collection and Enforcement) (Local List) Regulations 1989, reg 22; Non-Domestic Rating (Collection and Enforcement) (Central List) Regulations 1989, reg 10.

7.11   Any such claim for a repayment is an 'action to recover any sum recoverable by virtue of any enactment' so that a limitation period of six years applies.[11] This means that an action to recover the money must be brought within six years of the date when the cause of action 'accrues'. It seems that the cause of action will accrue as soon as regulation 9 applies and there has been an overpayment. It may be arguable that the cause of action will not 'accrue' until the ratepayer has requested a repayment. This argument is unlikely to be accepted, however, as there is provision in the regulations for money to be repaid (or credited to his account) even if the ratepayer has not requested it. As such it would seem that the cause of action accrues as soon as the entitlement arises.

7.12   This limitation period is strict and there is no provision to extend it, even if the ratepayer was unaware of his entitlement to the refund.

Interest

7.13   Regulations make provision for the addition of interest to repayments or credits made under the collection and enforcement regulations.[12] The entitlement to interest arises only when the repayment is the result of a list alteration.[13]

7.14   There is also a general power for the courts to award interest if proceedings are taken to recover money due.[14] This power applies to any claim for 'the recovery of a debt'. These words are almost certainly wide enough to cover a claim for a repayment of overpaid rates, however brought.[15] The power to award interest is discretionary, although a

---

11 Limitation Act 1980, s9(1).

12 The Non-Domestic Rating (Payment of Interest) Regulations 1990.

13 The Non-Domestic Rating (Payment of Interest) Regulations 1990, reg 3(1).

14 If the claim is brought in the County Court, the relevant section is s69 of the County Courts Act 1984; in the High Court the relevant section is s35A of the Senior Courts Act 1981.

15 See *R (Kemp) v Denbighshire Local Health Board* [2006] EWHC 181 (Admin) at [86]: '"debt" in section 35A extends so as to cover sums of money subject to an obligation, however arising, to repay them'. This is consistent with the view of the House of Lords on the predecessor provisions in *DP Exploration Co (Libya) Ltd v Hunt (No 2)* [1983] 2 AC 352 at p373: the words 'debt or damages' were held to 'cover any sum of money which is recoverable by one party from another, either at common law or in equity or under a statute'.

court will usually award interest at a rate sufficient to account for: (i) inflation; and (ii) the fact that the claimant has been kept out of money rightfully due to him.

RESTITUTION OF UNLAWFUL TAXES

7.15 Where the regulations do not provide a right to recover an overpayment, the common law may do so. The common law recognises a general right to restitution of money paid pursuant to an unlawful levy of tax.[16] This extends to payments of rates.[17] There is no requirement that the tax has been paid pursuant to a demand.[18]

7.16 This principle may in particular assist a ratepayer in two situations where the regulations are silent. The first is where an overpayment is made even though there has been no demand for the amount in question. The regulations require a demand, but no demand is required for a claim on common law principles of restitution.

7.17 The second situation is where demands have been made which are apparently valid and which correctly calculate liability to pay, but which are served late and are therefore unenforceable.[19] The regulations will not assist in this situation because the amount in such demands does not exceed the liability calculated under LGFA 1988. One course for a ratepayer served with such demands is to resist payment in the Magistrates' Court on the basis that the demands are invalid. However, even if a ratepayer has failed to do this and has paid the amounts demanded he will still arguably be entitled to restitution of the money under common law principles.[20]

7.18 As explained above, interest is likely to be recoverable under the court's general powers.[21]

---

16 *Woolwich Equitable Building Society v IRC* [1993] AC 70 was the first case to recognise this right. See Lord Goff at p177F for a statement of the principle.

17 *R v Barking and Dagenham LBC ex p Magon* [2004] RA 269.

18 *Test Claimants in the FII Group Litigation v Revenue and Customs Commissioners* [2012] UKSC 19 at [79].

19 For late demands, see paras 5.4–5.

20 Liability orders had been made in *R v Barking and Dagenham LBC ex p Magon* [2004] RA 269 but the Court of Appeal nonetheless suggested that the ratepayer had a remedy in restitution.

21 See para 7.14.

Defences

7.19   The defences to a restitutionary claim of this kind are not clearly
       defined. The one defence which has been clearly recognised in the
       case law[22] is that money will not be recoverable where it has been paid
       to 'close the transaction'. This refers to a situation where the money
       is paid either in order to avert threatened litigation or deliberately
       in order to bring matters to a conclusion ('close the transaction').[23]
       Ordinarily, therefore, a ratepayer who has failed to dispute the
       demands in the Magistrates' Court will be precluded from recovering
       back the money he paid because he will be held to have paid to close
       the transaction. It is a question of fact in every case, however, whether
       this is indeed what he did, and if payment was made under protest or
       'under the compulsion of urgent and pressing necessity'[24] then that
       may mean that the cause of action is still available.

Limitation

7.20   A six-year time period applies to a cause of action under this com-
       mon law principle. The reason why a six-year time period applies
       is not clear. It is either because a restitutionary claim is a claim in
       'quasi-contract' such that the six-year limitation period for contrac-
       tual claims[25] applies or a claim in tort such that the six-year limitation
       period for tort claims[26] applies. It appears to be well established that
       one or the other would apply.[27]

---

22 *Woolwich Equitable Building Society v IRC* [1993] AC 70 per Glidewell LJ at p98,
    Butler-Sloss LJ agreeing. *Test Claimants in the FII Group Litigation v Revenue and
    Customs Commissioners* [2012] UKSC 19 at [79].
23 *Woolwich Equitable Building Society v IRC* [1993] AC 70 per Glidewell LJ at p98.
24 *Maskell v Horner* [1915] 3 KB 106 at p118.
25 Limitation Act 1980, s5.
26 Limitation Act 1980, s2.
27 It was agreed that one or the other period would apply at first instance in *Deutsche
    Morgan Grenfell v IRC* [2003] EWHC 1866 (Ch) at [7] and not doubted on appeal,
    where the whole issue was one of limitation; see *Deutsche Morgan Grenfell v IRC*
    [2006] UKHL 49 at [50]. See also *Woolwich Equitable Building Society v IRC* [1993]
    AC 70 per Lord Goff at p174 who apparently considered that a six-year limitation
    period would apply, although without stating why.

RESTITUTION OF MONEY PAID UNDER A MISTAKE

7.21   The House of Lords has clarified in the case of *Deutsche Morgan Grenfell v IRC*[28] that there exists a further cause of action for repayment of taxes. This applies where the tax has been paid under a mistake, whether that mistake is one of fact or one of law. It was previously thought that money paid under a mistake of law was not recoverable.[29] This cause of action is available to a ratepayer.[30]

7.22   The advantages for a ratepayer of claiming pursuant to a mistake of fact or law are twofold. First, if it can be established that the rates were paid because of a mistake, the ratepayer is unlikely to be held to have paid to 'close the transaction'. He is more likely to have paid because he honestly believed the rates were due. He therefore avoids the main defence to a general restitutionary claim, as set out above. Second, where the claim is for money paid under a mistake, there is the possibility of extending the six-year limitation period.

7.23   It may be argued that the provisions in the regulations on the recovery of overpaid rates mean that a claim under this cause of action is excluded. Such an argument seems unlikely to succeed, however, given that the regulations do not allow any relief specifically from the consequences of a mistake.[31]

7.24   In order to establish such a claim, the ratepayer must establish three things: (1) he made a mistake; (2) the mistake caused the payment; and (3) the billing authority had no right to receive the payment.[32]

7.25   As to the first of these, the question of whether a mistake was made is a question of the ratepayer's actual understanding of the facts and the law – his 'state of mind'.[33] It does not matter if his mistaken

---

28   [2006] UKHL 49, [2007] 1 AC 558.
29   It appears to have been long established that overpayments of rates made under a mistake of fact are recoverable: *Meadows v Grand Junction Waterworks Co* (1905) 69 JP 255, which concerned water rates; and *Burland v Hull Local Board of Health* (1862) 3 B&S 271, where there was no dispute that the overpaid rates were recoverable in principle.
30   *Monster Worldwide v Westminster CC* [2009] RVR 186 is an example of it being used.
31   In *Deutsche Morgan Grenfell v IRC* [2006] UKHL 49 at [19], [52]–[55], a similar argument based on s33 of the Taxes Management Act 1970 (which does include provision for mistakes) was rejected.
32   *Kleinwort Benson v Lincoln CC* [1999] 2 AC 349 per Lord Hope at p407H.
33   *Deutsche Morgan Grenfell v IRC* [2006] UKHL 49 at [20].

understanding was unreasonable. The more unreasonable his view appears to have been, of course, the less likely it is to be accepted as his actual view. The concept of mistake includes cases of sheer ignorance as well as of positive but mistaken belief.[34]

7.26 It is also important to note that mistake of law can arise somewhat artificially where a later decision of the courts changes the law as it was at the time of the payment. In those circumstances, the ratepayer will be held to have been acting under a 'mistake of law', even though, at the time, his view of the law may have been apparently correct and shared by most people. In other words, it is no defence to such a claim that the ratepayer's mistaken view of the law was widely shared or was a settled view of the law.[35]

7.27 As to the second element of the claim, this can be demonstrated by asking whether the ratepayer would have made the payment if he had known the true facts or the true view of the law.[36] It is not enough that the ratepayer took a risk that he might be wrong about the law and paid anyway; it must be shown that the mistaken view caused the payment.

7.28 The third element is included because the foundation of this part of the law is the theory of unjust enrichment. Normally if a payment is made on a mistaken basis then it would be unjust for the recipient to be allowed to retain it. If there was an alternative and correct basis on which the billing authority was entitled to receive the money, however, then there is no injustice in allowing the billing authority to retain it. As such the money will not be recoverable by the ratepayer.

7.29 As explained above, interest is likely to be recoverable under the court's general powers.[37]

Defences

7.30 The defences to a claim in mistake of fact or law are the defences generally applicable to restitutionary claims:[38]

(1) Estoppel
(2) Change of position

---

34 *Kleinwort Benson v Lincoln CC* [1999] 2 AC 349 per Lord Hope at p410B.
35 *Deutsche Morgan Grenfell v IRC* [2006] UKHL 49 at [18], [23], [145].
36 See per Lord Hope in *Kleinwort Benson v Lincoln CC* [1999] 2 AC 349 at p411C and in *Deutsche Morgan Grenfell v IRC* [2006] UKHL 49 at [59]–[60].
37 See para 7.14.
38 *Kleinwort Benson v Lincoln CC* [1999] 2 AC 349 per Lord Hope at p412C.

(3) Money paid as or as part of a compromise

(4) Money paid in settlement of an honest claim.

7.31 Estoppel operates to prevent a person from going back on a previous assertion. It requires a clear assertion which leads the other party to change his position to his detriment. It is hard to envisage many situations in which the ratepayer would make such an assertion. Even if the ratepayer asserted that the rates were payable, it is hard to imagine how this could give rise to a detrimental change of position by the billing authority. This defence is likely to be of very limited application.

7.32 Change of position refers to a situation where the person receiving the money has subsequently changed his position detrimentally such that it would be unfair to require him to pay it back. An example often given is if the recipient has donated the money to charity. As has been observed, this defence will rarely if ever apply to wrongly extracted tax.[39] It seems that it will only apply if the money has been spent and the mistaken payment is entirely the fault of the ratepayer.[40]

7.33 Money paid as or as part of a compromise represents a similar defence to that of money paid to 'close the transaction', discussed above.[41] However, it would appear that something more is required – namely, something that can be described as a real compromise of a claim. This defence is sometimes treated as being simply a denial that there has been any mistake in the first place; it is the prime example of a situation in which someone took the risk that the payment was not due but decided to pay anyway in order to conclude matters.[42]

7.34 The final defence – that of money paid in settlement of an honest claim – has been described as 'somewhat undefined'.[43] In a sense, any money paid in response to a demand notice issued in good faith will have been paid in settlement of an honest claim. Such a broad

---

39 *Deutsche Morgan Grenfell v IRC* [2006] UKHL 49 per Lord Walker at [145].

40 *Larner v LCC* [1949] 2 KB 683 at p688 suggests that some fault on the part of the payee is required before the defence will apply, although it was not decided in the context of repayments of tax. *Spiers and Pond Ltd v Finsbury MBC* (1956) 1 RRC 219 at p225 suggests that the defence will apply whenever the money has been spent; this would appear to be incorrect in the light of *Larner* and later decisions of the higher courts.

41 See para 7.19.

42 *Deutsche Morgan Grenfell v IRC* [2006] UKHL 49 at [27], [65].

43 *Kleinwort Benson v Lincoln CC* [1999] 2 AC 349 per Lord Goff at p385B.

argument is unlikely to succeed, however. Again it appears that there needs to have been a voluntary acceptance of risk that the payment might not be due.[44] The scope of this defence will need to be clarified in future decisions.

7.35   It will be clear that these defences are potentially of rather limited application. This fact, together with the ability to extend the limitation period, may make this cause of action attractive to ratepayers seeking a refund of rates paid.

Limitation

7.36   The standard limitation period of six years applies, on the same basis as is explained above.[45] Ordinarily, time begins to run from the date the payment was made.[46] As the action is for relief from the consequences of a mistake, however, the six-year period will not in fact begin to run 'until the [ratepayer] has discovered the fraud, concealment or mistake (as the case may be) or could with reasonable diligence have discovered it'.[47]

7.37   It will be question of fact when the mistake was actually discovered, to be dealt with by evidence from the ratepayer. As noted above, it is possible to make a 'mistake' of law by following an understanding of the law which is later held to have been incorrect. In such a case, the mistake will normally be treated as having been discovered when the ratepayer becomes aware of the judgment clarifying or changing the law.[48] It is still a question of fact when the mistake was discovered, however, and there may be cases 'where a party may be held to have discovered a mistake without there being an authoritative pronouncement directly on point on the facts of that case by a court'.[49] It is at least arguable that the mistake is discovered as soon as the ratepayer is aware he has a worthwhile claim.[50]

---

44   *Kleinwort Benson v Lincoln CC* [1999] 2 AC 349 per Lord Hope at p413D.

45   See para 7.20.

46   *Kleinwort Benson v Lincoln CC* [1999] 2 AC 349 at p386F.

47   Limitation Act 1980, s32(1).

48   *Deutsche Morgan Grenfell v IRC* [2006] UKHL 49 at [31], [71].

49   Words cited with approval by Lord Walker in *Deutsche Morgan Grenfell v IRC* [2006] UKHL 49 at [144].

50   *Deutsche Morgan Grenfell v IRC* [2006] UKHL 49 per Lord Brown at [165], who was however 'in a minority of one' on this issue.

7.38 Even if the mistake has not in fact been discovered, it is possible to argue that time has started to run because it could have been discovered 'with reasonable diligence'. 'Reasonable diligence' does not impose a requirement to do everything possible, only to do what an ordinary prudent person would do having regard to all the circumstances.[51] Another way of putting it is that the burden of proof is on the ratepayer to establish that the mistake could not have been discovered without exceptional measures which he could not have been expected to take. The test is to ask how a person carrying on a business of the relevant kind would act if he had adequate but not unlimited staff and resources.[52] Whether or not this test is satisfied will be a question of fact.

---

51 *Peco Arts Inc v Hazlitt Gallery Ltd* [1983] 1 WLR 1315.
52 As it was put by the Court of Appeal in *Paragon Finance plc v D.B. Thakerar (A Firm)* [1999] 1 All ER 400 at p418.

Part II

Council tax

8

Chargeable dwellings

8.1 Historically, domestic property was subject to rates. The enactment of LGFA 1988, as well as providing the modern framework for rating, also removed domestic property from the rating system, and created the community charge (better known as the 'poll tax'). This marked a radical departure from the historic principle of taxation based on the value of the property. The swift demise of the poll tax was followed by the creation of the council tax in LGFA 1992, and a return to something much more akin to the rating system, although with significant modifications. This chapter deals with the concept of the 'chargeable dwelling', which is the unit of taxable property for council tax purposes.

8.2 The interaction of the council tax provisions with the provisions on non-domestic rates means that it is sometimes necessary to refer back to the rating system in order to understand the council tax system. References are provided where appropriate to the earlier part of this book, on non-domestic rating.

8.3 A chargeable dwelling is defined as any dwelling which is not an exempt dwelling.[1] This chapter first considers how a 'dwelling' is to be identified, then looks at which dwellings will be exempt. Finally there is a brief section on valuation lists and the valuation of dwellings for council tax purposes.

DWELLINGS

8.4 Council tax is, fundamentally, a tax payable in respect of dwellings.[2] A 'dwelling' is defined by reference to the definition of a hereditament.

1 LGFA 1992, s4(1)–(2).
2 LGFA 1992, s1(1).

In essence, it is property which meets the definition of a hereditament for rating purposes, but which is not required to be shown in a rating list and is not exempt from rating.[3] The practical effect of this definition is that a dwelling equates with a hereditament which is domestic property.[4] The definition of domestic property is reproduced here for ease of reference:

> property is domestic if:
>
> (a) it is used wholly for the purposes of living accommodation,
> (b) it is a yard, garden, outhouse or other appurtenance belonging to or enjoyed with property falling within paragraph (a) above,
> (c) it is a private garage which either has a floor area of 25 square metres or less or is used wholly or mainly for the accommodation of a private motor vehicle, or
> (d) it is private storage premises used wholly or mainly for the storage of articles of domestic use.[5]

8.5 A composite hereditament, that is, one which is partly domestic and partly non-domestic, is also a dwelling as long as it includes property which falls within sub-paragraph (a) of the definition of domestic property.[6] The 'domestic' part does not need to be a self-contained dwelling or to be capable of separate occupation; it is enough if there is one room which is used wholly for the purposes of living accommodation.[7]

8.6 In the same way, property which falls within sub-paragraphs (b), (c) and (d) will not be a dwelling unless it forms part of a larger property which includes property within sub-paragraph (a).[8] In England, domestic electricity-generation installations can be domestic property, but again they will not be a 'dwelling' except insofar as they form part of a larger property which includes property within sub-paragraph (a).

---

3 LGFA 1992, s3(2).
4 On the definition of hereditaments, see para 2.2 onwards. On domestic property, see para 2.23 onwards.
5 LGFA 1988, s66(1).
6 LGFA 1992, s3(3).
7 *Williams v Bristol District VO* [1995] RA 189 at p194–195.
8 LGFA 1992, s3(4).

8.7   The normal process for identifying a dwelling is therefore as follows:

(1) identify the hereditament;
(2) identify whether it or any part of it is 'used wholly for the purposes of living accommodation';
(3) if so it is a dwelling.

8.8   It should be remembered that there are specific rules for property used for various kinds of short-stay accommodation, which may not be domestic or a dwelling even if it is used wholly for the purpose of living accommodation.[9]

8.9   When applying these tests, 'it shall be assumed that any state of affairs subsisting at the end of the day had subsisted throughout the day'.[10] It is submitted that this provision does not require a focus on the situation at 11.59 p.m. The following remarks on the parallel provision in LGFA 1988 are relevant here:

> where s67(5) refers to the state of affairs existing immediately before the day ends it is not requiring that attention be confined to the particular activities being carried on at a precise moment in time. What has to be considered is the use of the property with all its features, and all that s67(5) does is to identify the material time by reference to which any change in the use of the property is to be related.[11]

8.10   The normal principles applicable to the identification of hereditaments (and, therefore, dwellings) are modified in certain cases, which are discussed below.

Disaggregation into separate dwellings

8.11   The Secretary of State is given the following powers under LGFA 1992:

> The Secretary of State may by order provide that in such cases as may be prescribed by or determined under the order—

---

9 See para 2.39 onwards for a full discussion of the rules applicable to short-stay accommodation.
10 LGFA 1992, s2(2).
11 *Tully v Jorgensen (VO)* [2003] RA 233 at p241–242.

(a) anything which would (apart from the order) be one dwelling shall be treated as two or more dwellings; and (b) anything which would (apart from the order) be two or more dwellings shall be treated as one dwelling.

8.12 These powers have been exercised in the Council Tax (Chargeable Dwellings) Order 1992. This concerns provisions on both 'disaggregation' (i.e. treating what would normally be one dwelling as more than one dwelling) and 'aggregation' (i.e. treating what would normally be more than one dwelling as one dwelling). The effect of the order, where it applies, is thus to split up or combine dwelling(s) identified on normal principles. It is therefore very important, before considering the effect of the order, to establish first what the hereditaments/dwellings would be on normal principles.[12] If this is not done, then the decision will be legally flawed.[13]

8.13 The main provisions on disaggregation are as follows (there are also specific provisions on care homes and refuges which are discussed below):

'single property' means property which would, apart from this Order, be one dwelling within the meaning of section 3 of [LGFA 1992];

'self-contained unit' means a building or a part of a building which has been constructed or adapted for use as separate living accommodation.[14]

where a single property contains more than one self-contained unit . . . the property shall be treated as comprising as many dwellings as there are such units included in it and each such unit shall be treated as a dwelling.[15]

8.14 It is therefore necessary to identify whether the property contains more than one 'self-contained unit'. If it does, each such unit will be treated as a separate dwelling even though they are all occupied together as one hereditament.

---

12 These are discussed in detail at para 2.2 onwards.
13 *R (Listing Officer) v Callear* [2012] EWHC 2697 (Admin) at [23] applying *Rawsthorne v Parr* [2009] EWHC 2002 (Admin) at [21]–[22].
14 Council Tax (Chargeable Dwellings) Order 1992, art 2.
15 Council Tax (Chargeable Dwellings) Order 1992, art 3.

8.15 There is a large number of decisions on the identification of self-contained units. They have been helpfully summarised by the High Court[16] as follows:

(1) The question is whether the effect of the construction or adaptation is such as to make the relevant building or part of a building reasonably suitable for use as separate living accommodation. I prefer the expression 'reasonably suitable' for such use to 'capable' of such use, because it makes clear that what matters is its fitness for that purpose by reference to contemporary standards of what is reasonable, not merely whether it might conceivably be used for such purpose however remote the possibility.

(2) The question is to be answered by reference to the physical characteristics of the building. This is sometimes referred to as a 'bricks and mortar test', but the epithet does not accurately capture the wide range of physical characteristics which may be of relevance including services and fixtures.

(3) This is an objective test. The test is not concerned with when, how or why those characteristics were achieved. The purpose of the construction or adaptation is irrelevant. The test is addressed to the result of the building work, not the circumstances in which it was carried out. Intention is irrelevant.

(4) Whether the test is met is a matter of fact and degree for the tribunal.

(5) There is a divergence in the authorities as to whether the actual use to which the building has been or is being put is capable of being a relevant consideration. The decision attaches to the dwelling through changes of ownership and use, such that actual use at any given time will rarely help to inform the result of applying an objective test based on the physical characteristics of the building. I prefer the view that actual use may in some cases be of some relevance. If, for example, the part of the property has in fact been used, or is being used, for occupation by persons who do not form part of a single household with those who occupy the remainder of the property, that may be a factor which supports a conclusion that its

---

16 In *Corkish (LO) v Wright* [2014] EWHC 237 (Admin) at [5].

physical characteristics make it suitable for such occupation. However actual use is not the test, and even in cases where it may be of some relevance it will not usually be a factor of significant weight. At most it may reinforce a decision reached by reference to the physical characteristics of the building.

(6) If what is being considered is part of a building, the physical characteristics to be considered include those of the remainder of the building as well as the part being considered. Access is one aspect of such characteristics. Separate public access may be a pointer to the part being separate living accommodation; whereas if access is through the remainder of the building this may tell against the part being separate living accommodation. In the latter case different weight may be attached where access is through the living areas of the remainder of the building from the weight to be attached where it is through a hallway. But access is not a factor which can be determinative without considering the other physical aspects of the building. The weight to be attached to it is a matter for the tribunal.

8.16   These are the principles which need to be applied. A few other relevant cases can be referred to in order to illustrate the application of these principles and highlight some areas which still require clarification.[17]

8.17   The first principle requires regard to be had to the unit's 'fitness for that purpose by reference to contemporary standards of what is reasonable'. This includes the need for 'an objectively acceptable degree of privacy'.[18] A door is not necessary; this standard of privacy may be provided in other ways.[19]

---

17   The full list of relevant cases is as follows: *Batty v Burfoot etc.* [1995] RA 299, *Butterfield v Ulm* (1997) 73 P&CR 289, *Beasley (LO) v National Council of YMCAs* [2000] RA 429, *McColl v Subacchi (LO)* [2001] EWHC 712 (Admin), *Clement (LO) v Bryant* [2003] EWHC 422 (Admin), *Coleman (LO) v Rotsztein* [2003] RA 152, *Williams (LO) v RNIB* [2003] EWHC 1308 (Admin), *Daniels (LO) v Aristides* [2006] EWHC 3052 (Admin), *Jorgenson (LO) v Gomperts* [2006] EWHC 1885 (Admin), *R (Listing Officer) v Callear* [2012] EWHC 2697 (Admin), *Corkish (LO) v Wright* [2014] EWHC 237 (Admin), *Kelderman v VOA* [2014] EWHC 1592 (Admin), *Ramdhun v VTE* [2014] EWHC 946 (Admin), *Kelderman v VOA* [2014] EWHC 1592 (Admin).

18   *Jorgenson (LO) v Gomperts* [2006] EWHC 1885 (Admin) at [27].

19   *Jorgenson (LO) v Gomperts* [2006] EWHC 1885 (Admin), *Ramdhun v VTE* [2014] EWHC 946 (Admin).

8.18   In terms of the fourth principle, the determination is a question of fact for the Valuation Tribunal for England (VTE). The High Court will not interfere with that determination unless there is an error of law. In some cases, however, the facts are so clear that the High Court has held that there is only one possible outcome that the VTE could reach. In the first, the court was dealing with flats. Each flat had a sink, cooker, fridge, shower, wash basin, toilet and lockable door. There were some communal facilities in the building. The only acceptable conclusion was that they were self-contained.[20] In the second case, the court was dealing with bedsits which had cooking facilities but no bathing facilities. The only conclusion could be that they were self-contained units.[21]

8.19   The fifth principle records the debate in the cases about the relevance of actual use. For the same reasons, there is doubt as to whether the terms of any planning permission will be relevant in their own right rather than as a matter which throws light on other relevant considerations.[22] Furthermore, the fact that the unit is not practically saleable as a separate unit is irrelevant; but if it is practically saleable as a separate unit that points towards a finding that it is self-contained.[23]

8.20   The sixth principle requires consideration of the physical characteristics of the wider building. Where these disclose the purpose for which the building as a whole has been constructed or adapted, that purpose will be relevant to a consideration of whether parts of the building are self-contained units or not.[24]

8.21   A registered care home[25] is treated differently in terms of disaggregation. It is to be treated as 'comprising the number of dwellings found by adding one to the number of self-contained units occupied by,

---

20 *Beasley (LO) v National Council of YMCAs* [2000] RA 429.

21 *Clement (LO) v Bryant* [2003] EWHC 422 (Admin). See also *R (Listing Officer) v Callear* [2012] EWHC 2697 (Admin), in which the court expressed the view that a bedsit with shower and cooking facilities but no toilet was clearly a self-contained unit.

22 *Batty v Burfoot etc.* [1995] RA 299.

23 *Batty v Burfoot etc.* [1995] RA 299 at p310.

24 *Williams (LO) v RNIB* [2003] EWHC 1308 (Admin), where the fact that the building was a home for blind people was treated as a factor weighing against a decision that the units in it were self-contained.

25 See Council Tax (Chargeable Dwellings) Order 1992, art 2 for the full definition.

or if currently unoccupied, provided for the purpose of accommodating the person registered . . . and each such unit shall be treated as a dwelling'.[26] Thus where a care home contains more than one unit of self-contained staff accommodation, that property does not fall to be entered as a separate dwelling.

8.22 In Wales, a refuge which is run otherwise than for profit to provide temporary accommodation to the victims of abuse[27] is to be treated as one dwelling.[28]

Aggregation of separate dwellings

8.23 In addition to the provisions dealing with the disaggregation of single properties, provision has also been made to aggregate multiple properties into one dwelling. The key parts of the Council Tax (Chargeable Dwellings) Order 1992 are as follows:

> 'multiple property' means property which would, apart from this Order, be two or more dwellings within the meaning of section 3 of [LGFA 1992];
>
> 'self-contained unit' means a building or a part of a building which has been constructed or adapted for use as separate living accommodation.[29]
>
> 4 (1) Where a multiple property—
>
> (a) consists of a single self-contained unit, or such a unit together with or containing premises constructed or adapted for non-domestic purposes; and
>
> (b) is occupied as more than one unit of separate living accommodation.
>
> the listing officer, may, if he thinks fit, subject to paragraph (2) below, treat the property as one dwelling.
>
> (2) In exercising his discretion in paragraph (1) above, the listing officer shall have regard to all the circumstances of the case, including the extent, if any, to which the parts of the property separately occupied have been structurally altered.

---

26 Council Tax (Chargeable Dwellings) Order 1992, art 3A.
27 Council Tax (Chargeable Dwellings) Order 1992, art 2 for full definition.
28 Council Tax (Chargeable Dwellings) Order 1992, art 3B.
29 Council Tax (Chargeable Dwellings) Order 1992, art 2.

8.24 These provisions will accordingly only apply where there is a single 'self-contained unit' which falls to be treated as more than one dwelling/hereditament on normal principles. In other words, a building or part of a building which has been constructed or adapted for use as separate living accommodation which is being occupied as separate 'hereditaments'.

8.25 The most common example of such a property would be a house in which each habitable room has been let out under a separate tenancy agreement, with the bathroom and kitchen being shared among all the occupiers. Each room will in these circumstances be a separate dwelling on normal principles, because it is subject to separate rateable occupation.

8.26 Where such a property exists, the listing officer has a discretion. He may, if he thinks fit, treat it as one dwelling. The Order specifies that one circumstance of particular relevance is 'the extent, if any, to which the parts of the property separately occupied have been structurally altered'.

8.27 Where there is one obvious self-contained unit (the original house in the example given above), this provision will not be difficult to apply. Where, however, multiple floors or parts of a building in use as separate dwellings would pass the test for a self-contained unit it will not be easy to see how the provisions should operate. It seems sensible to conclude that the whole building cannot in those circumstances be a 'self-contained unit',[30] and that the search should be for the most natural division of the building into self-contained units, each of which can then be aggregated if appropriate.

8.28 The relevant factors for the listing officer's discretion include the following (as well as the extent of any structural alterations): (a) the degree of sharing or common facilities; (b) the degree of adaptations to the property; (c) whether separate units of accommodation can be clearly identified; and (d) the degree of transience of the occupiers.[31] A low degree of alteration/adaptation and a high degree of sharing and/or transience all suggest that the listing officer should exercise his discretion.

---

30 The parties and the court in *McColl v Subacchi (LO)* [2001] EWHC 712 (Admin) all appear to have agreed that where one part of a larger building was a 'self-contained unit', this implied that the remaining part was also to be considered such a unit.

31 These factors were listed as relevant with apparent approval in *James v Williams (VO)* [1973] RA 305 at p307.

8.29   It appears that no appeal to the VTE is possible in respect of an aggre-
       gation decision by the listing officer.[32] Instead, such a decision must be
       challenged by way of judicial review proceedings.[33] This may have to
       be pursued in tandem with an appeal to the VTE where the taxpayer
       is also challenging the listing officer's approach to the identification of
       the relevant dwellings.

Incomplete dwellings

8.30   As has been explained with reference to the rating system, property
       which is not yet ready for occupation is not a hereditament.[34] It follows
       that a 'dwelling' which is not yet ready for occupation is not a dwelling
       and does not give rise to any council tax liability.[35]

8.31   As with the rating system, however, a billing authority is able to serve
       a notice known as a 'completion notice', which effectively deems that
       a dwelling is complete and ready for occupation.

8.32   LGFA 1992 takes the approach of applying the completion notice
       regime for the purposes of non-domestic rates to dwellings, subject
       to a number of small modifications. The modifications essentially do
       no more than adapt the completion notice regime so that it applies to
       dwellings. They can be summarised as follows:

       (1)   The completion notice regime for dwellings applies to new
             buildings, as does the rating regime. In the case of dwellings,
             this includes a building produced by the structural alteration
             of an existing building. This is so whether or not the exist-
             ing building is made up of any dwelling or dwellings.[36] The
             regime for dwellings would therefore seem to apply whenever

---

32   Council Tax (Alteration of Lists and Appeals) (England) Regulations 2009, reg
     4(1)(a) specifically excludes them; it is not clear whether reg 4(6) would in
     appropriate circumstances provide a broader right to challenge an aggregation
     decision.

33   *R v London South East Valuation Tribunal and Neale, ex parte Moore* [2001] RVR 92
     is sometimes cited as deciding otherwise, but it is: (1) not clear on the point; and
     (2) not a binding authority (being a decision on an application for permission to
     appeal only).

34   The relevant principles are fully discussed in Chapter 6 at para 6.3 onwards.

35   See *RGM Properties Ltd v Speight (LO)* [2011] EWHC 2125 (Admin) for an
     application of this test.

36   LGFA 1992, s17(6).

a structural alteration appears to produce a new dwelling or dwellings;

(2) The effect of a council tax completion notice is that 'any dwelling in which the building or any part of it will be comprised shall be deemed . . . to have come into existence' on the relevant day;[37]

(3) The 'relevant day' is either the day stated in the notice or, if an appeal is brought, the day as determined in accordance with LGFA 1988, schedule 4A.[38] This appears to suggest that the power to agree a date in writing, or to withdraw the notice, are only triggered if and when an appeal has been brought;

(4) If the notice relates to a building produced by the structural alteration of a building containing one or more dwellings, those dwellings will also cease to exist on the relevant day;[39]

(5) Appeal against the decision of the VTE/VTW is to the High Court, not to the Upper Tribunal (Lands Chamber), and must be made within four weeks.[40]

8.33 There is no modification of the law on the service, validity and withdrawal of completion notices, or on appeals against completion notices, which applies to council tax completion notices as it does to those relating to non-domestic hereditaments.

Disrepair

8.34 A dwelling which has fallen into dereliction may therefore cease to be a dwelling altogether. The question is whether the property is capable of being rendered suitable for occupation as a dwelling by a reasonable amount of repair works. The distinction is between a truly derelict property, which is incapable of being repaired, and a property which is merely in disrepair and which could be repaired to make it capable of occupation as a dwelling. Whether the repairs would be economic or uneconomic to undertake is not a crucial distinction.[41] The cost of repairs required may be an indication of

---

37 LGFA 1992, s17(3).
38 LGFA 1992, s17(4).
39 LGFA 1992, s17(5).
40 Valuation Tribunal for England (Council Tax and Rating Appeals) (Procedure) Regulations 2009, reg 43(1); Valuation Tribunal for Wales Regulations 2010, reg 44.
41 *Wilson v Coll (LO)* [2011] EWHC 2824 Admin at [40]–[41].

whether they are reasonable repairs on the one hand or works of reconstruction on the other.

EXEMPT DWELLINGS

8.35   If a dwelling is an exempt dwelling, it will not be a chargeable dwelling and therefore no liability to pay council tax will arise in respect of it.[42] Billing authorities are responsible for determining whether a dwelling is exempt or not and there is a right of appeal to the VTE against a decision of the billing authority.[43]

8.36   The classes of exempt dwelling are set out in the Council Tax (Exempt Dwellings) Order 1992. In order to be exempt, a dwelling must fall within one of the classes prescribed. Each class is designated by a letter. The order has been heavily amended over the years, with classes being added and removed, such that some letters no longer have a corresponding class. Those classes which are currently in force are dealt with in turn below. Three points of interpretation recur throughout and are dealt with here.

8.37   First, when calculating continuous periods of vacancy or periods during which a dwelling has been unoccupied, any break in the period of six weeks or less is to be disregarded.[44] 'Unoccupied' in this context means a dwelling in which no one lives.[45] This prevents a short period of occupation or use from triggering a further period of exemption.

8.38   Second, various classes of exemption refer to a dwelling which is the 'sole or main residence' of a person. This is a key concept in determining liability to council tax, and is dealt with in this book in the chapter on liability.[46]

8.39   Third, various classes also make reference to the owner or tenant of a dwelling. 'Owner' is defined in the same way as in relation to liability.[47] 'Tenant' is given an expansive definition which essentially covers anyone with a contractual right to occupy the dwelling. Tenant in this context means a person who:

---

42   LGFA 1992, s4.
43   LGFA 1992, s16. Appeals and challenges generally are discussed at para 9.60 onwards.
44   Council Tax (Exempt Dwellings) Order 1992, art 2(2)–(3).
45   Council Tax (Exempt Dwellings) Order 1992, art 2.
46   See paras 9.10–14.
47   On which, see LGFA 1992 s6(5), discussed at paras 9.8–9.

    (1) has a leasehold interest in a dwelling which was granted for a term of less than six months;

    (2) is a secure, introductory or statutory tenant of a dwelling; or

    (3) has a contractual licence to occupy a dwelling.[48]

8.40   Finally, it should be noted that Classes D, E, I and J overlap. All are concerned with owners or tenants who are absent for various specified reasons. If someone is absent for a variety of these reasons in succession, the dwelling which was formerly his sole or main residence will continue to be exempt, notwithstanding that the reason for his absence has changed over time.[49]

Class A: major repair work

8.41   In Wales only, a dwelling will be exempt for 12 months if it is vacant and one of three requirements apply. The first is that it requires or is undergoing major repair works to make it habitable; habitable in this context is a lower standard than 'marketable' or 'lettable'.[50] The second is that it is undergoing structural alteration. The third is that one or other of these types of work have been substantially completed and the dwelling has been vacant since, for a period of no more than six months.

8.42   In order to be considered vacant, the dwelling must be unoccupied. If it consists of a pitch or mooring it will be vacant if the associated caravan or boat is itself unoccupied. In any other case, the dwelling must also be substantially unfurnished. Periods during which the dwelling is not vacant which are shorter than six weeks are to be disregarded.[51] This prevents the owner from incurring repeated periods of exemption by, for example, returning the furniture for a short period. There would appear to be nothing to prevent an owner from securing a further period of exemption for a property on which works were required to be done by simply returning the furniture for six weeks and one day, however.

---

48 Council Tax (Exempt Dwellings) Order 1992, art 2.

49 This is the effect of the definition of 'relevant absentee' in Council Tax (Exempt Dwellings) Order 1992, art 2.

50 *Edem v Basingstoke and Deane BC* [2012] EWHC 2433 (Admin).

51 Council Tax (Exempt Dwellings) Order 1992, art 2(2).

Class B: charitable exemption

8.43   A dwelling will be exempt if it satisfies the following four requirements:

> (i) it is owned by a body (ii) that body is established for charitable purposes only (iii) the dwelling is unoccupied and has been so for a period of less than 6 months and (iv) it was last occupied in furtherance of the objects of the charity.

8.44   All of these requirements must be satisfied before the dwelling will be exempt. There can be no presumption that any requirement is fulfilled in a particular case, even where the property is owned by a provider of social housing. However, a short statement addressing the four requirements and stating that they are met will usually be sufficient proof that they are met.[52]

8.45   As with Class A exemption, a period of occupation which lasts for less than six weeks is to be discounted when applying this exemption.[53] This prevents a short period of occupation from triggering a further six-month period of exemption.

8.46   It should be noted that although this class applies to both England and Wales it is only really of relevance in England. In Wales, Class C remains in force such that there is no need to rely on Class B.

Class C: unoccupied dwellings

8.47   In Wales only, a dwelling is exempt if it has been unoccupied for less than six months. As with Class A exemption, a period of occupation which lasts for less than six weeks is to be discounted when applying this exemption.[54] This prevents a short period of occupation from triggering a further six-month period of exemption.

Class D: detainees

8.48   A dwelling is exempt where it would be the sole or main residence of its owner or tenant, but that owner or tenant is in prison or detained for mental-health reasons.[55] Similarly, if it was the sole

---

52 *Ealing LBC v Notting Hill Housing Trust* [2015] EWHC 161 (Admin) at [19].
53 Council Tax (Exempt Dwellings) Order 1992, art 2(3).
54 Council Tax (Exempt Dwellings) Order 1992, art 2(3).
55 See LGFA 1992 sch 1 para 1 for the full list of circumstances.

or main residence of such a person before his detention and he has remained absent since.

Class E: residents in care

8.49   An unoccupied dwelling is exempt where it was previously the sole or main residence of its owner or tenant and where that person has been resident in a hospital, nursing home or care home since ceasing to live at the dwelling.

Class F: deceased owner or tenant

8.50   An unoccupied dwelling is exempt from council tax where it has been unoccupied since the death of its owner or tenant. It will remain exempt for up to six months following the grant of probate or letters of administration if the person otherwise liable for council tax would be the executor or administrator.

Class G: occupation prohibited by law

8.51   An unoccupied dwelling is exempt where occupation is prohibited by law or where it is kept unoccupied by reason of action taken under statutory powers with a view to prohibiting its occupation or to acquiring it. These provisions are very similar to the exemption for certain unoccupied non-domestic property, and it seems they will be interpreted in the same way.[56]

8.52   In England, there is a further category of exemption under this class where occupation is prohibited under a planning condition. This responds to the fact that use in breach of planning control is not use 'prohibited by law' until and unless enforcement action is taken.[57]

Class H: ministers of religion

8.53   The following will be exempt:

> an unoccupied dwelling which is held for the purpose of being available for occupation by a minister of any religious denomination as a residence from which to perform the duties of his office.

56 See discussion of those provisions at paras 4.123–124.
57 See *Westminster CC v Regent Lion Properties Ltd* [1990] RA 121.

Class I: owner or tenant receiving care

8.54   This class extends the exemption provided by Class E. It applies to any unoccupied dwelling which was previously the sole or main residence of its owner or tenant, where that person now has his sole or main residence elsewhere for the purpose of receiving personal care required by reason of 'old age, disablement, illness, past or present alcohol or drug dependence or past or present mental disorder'. This avoids such people being denied the benefit of an exemption purely because they are not resident in a hospital etc. as defined in Class E.

Class J: owner or tenant providing care

8.55   An unoccupied dwelling is exempt where its owner or tenant is now resident elsewhere for the purpose of 'providing, or better providing' personal care for another who requires that care by reason of 'old age, disablement, illness, past or present alcohol or drug dependence or past or present mental disorder'.

Class K: students studying elsewhere

8.56   This class provides exemption for the dwellings of owners who are studying elsewhere. An unoccupied dwelling is exempt where the person who last had his sole or main residence there is a student, as long as any other liable owners are also students. 'Student' is given a particular definition, which includes student nurses, apprentices and youth training trainees.[58] In order to qualify as a 'student', a person must be enrolled on a full-time course of study which normally requires his attendance for at least 24 weeks in each year. This requirement is relatively flexible; the attendance requirement does not have to relate to a particular place, so a person studying from home could qualify.[59]

Class L: mortgagee in possession

8.57   An unoccupied dwelling is exempt where a mortgagee is in possession under the mortgage.

---

58 See LGFA 1992 sch 1 para 4 and the Council Tax (Discount Disregards) Order 1992 art 4 and sch 1.
59 *R (Feller) v Cambridge City Council* [2011] EWHC 1252 (Admin) at [67]–[68].

Class M: hall of residence

8.58 A hall of residence provided predominantly for the accommodation of students will be exempt provided one of two conditions is met. The first is that the owner or manager is a prescribed education establishment[60] or a body established for charitable purposes only. The second is that a prescribed educational establishment has power to nominate the majority of occupiers.

Class N: students

8.59 Dwellings occupied only by students are exempt. They remain exempt through the vacation even if the student occupiers have all moved out. Exemption is granted in similar way to dwellings which are occupied by the non-national spouses or dependents of students, who are present in the country simply to accompany the student, or which are occupied by school and college leavers[61] in the period between 30 April and 1 November in any year.

Class O: armed forces accommodation

8.60 A dwelling is exempt if it is owned by the Secretary of State for Defence and held for the purposes of armed forces accommodation. Class P deals with the position of visiting forces.

Class P: visiting forces

8.61 A dwelling is exempt if the person who would otherwise be liable is a member of a visiting force or its civilian component, or is a non-national dependent of such a person.[62]

Class Q: trustee in bankruptcy

8.62 An unoccupied dwelling is exempt if the person who would otherwise be liable to pay the council tax is a trustee in bankruptcy.

60 See LGFA 1992 sch 1 para 5 and the Council Tax (Discount Disregards) Order 1992 art 5(2) and sch 2.
61 See the Council Tax (Additional Provision for Discount Disregards) Regulations 1992, reg 3 Class C for the full definition of 'school and college leavers'.
62 For the full statement of those who qualify see the Visiting Forces Act 1952, s12(2), which defines the required 'relevant association'.

Class R: vacant pitches and moorings

8.63 A pitch or mooring which is not occupied by a caravan or boat, respectively, is exempt.

Class S: occupants aged under 18

8.64 A dwelling is exempt where all the occupants are aged under 18.

Class T: separate letting prohibited

8.65 A dwelling is exempt if it forms part of a single property with another dwelling and is not separately lettable otherwise than in breach of planning control. This provision mitigates the effects of the disaggregation provisions discussed above.[63]

Class U: mentally impaired persons

8.66 A dwelling is exempt if it is occupied by one or more severely mentally impaired persons. A severely mentally impaired person is someone who has been certified by a registered medical practitioner as having 'a severe impairment of intelligence and social functioning (however caused) which appears to be permanent'.[64]

Class V: diplomats

8.67 A dwelling is exempt if it is the only or main residence of a non-national benefiting from diplomatic privileges and immunities.

Class W: dependent relatives

8.68 This class provides exemption for what might commonly be referred to as 'granny flats' or 'granny annexes'. In so doing, it mitigates the impact of the provisions on disaggregation, which will normally require such accommodation to be treated as a separate dwelling. It applies to:

> a dwelling which forms part of a single property including at least one other dwelling and which is the sole or main residence of a dependent relative of a person who is resident in that other dwelling, or as the case may be, one of those other dwellings.

63 See para 8.11 onwards.
64 LGFA 1992, sch 1 para 2.

8.69 'Relative' in this context is given a full definition.[65] A relative is only a dependent relative for these purposes if he is over the age of 65 or is severely mentally impaired,[66] or substantially and permanently disabled.[67]

VALUATION LISTS

8.70 The listing officer for each billing authority is required to compile and maintain a list of dwellings.[68] It is to show the value of each dwelling, by reference to 'bands'. At present, these are as follows.[69] In England:

Band	Range of values
A	Values not exceeding £40,000
B	Values exceeding £40,000 but not exceeding £52,000
C	Values exceeding £52,000 but not exceeding £68,000
D	Values exceeding £68,000 but not exceeding £88,000
E	Values exceeding £88,000 but not exceeding £120,000
F	Values exceeding £120,000 but not exceeding £160,000
G	Values exceeding £160,000 but not exceeding £320,000
H	Values exceeding £320,000

8.71 In Wales:

Band	Range of values
A	Values not exceeding £44,000
B	Values exceeding £44,000 but not exceeding £65,000
C	Values exceeding £65,000 but not exceeding £91,000
D	Values exceeding £91,000 but not exceeding £123,000
E	Values exceeding £123,000 but not exceeding £162,000
F	Values exceeding £162,000 but not exceeding £223,000
G	Values exceeding £223,000 but not exceeding £324,000
H	Values exceeding £324,000 but not exceeding £424,000
I	Values exceeding £424,000

65 Council Tax (Exempt Dwellings) Order 1992, art 2(5).
66 See para 8.66.
67 Council Tax (Exempt Dwellings) Order 1992, art 2(4).
68 LGFA 1992, ss21, 23.
69 LGFA 1992, s5.

8.72   Valuation is by reference to 1 April 1991 values.[70] The value of any dwelling shall be taken to be the amount which it might reasonably have been expected to realise if it had been sold in the open market by a willing vendor on that date, on the basis of certain assumptions as to the condition of the dwelling and the sale.[71] The operation of those provisions is outside the scope of this book.

70   LGFA 1992, s21(2A).
71   See the Council Tax (Situation and Valuation of Dwellings) Regulations 1992, reg 6.

9

Liability to council tax

9.1 This chapter addresses the questions of *who* is liable and *how much* they are liable to pay. The basic principles as to the amount of liability are complicated by a large array of provisions allowing or requiring discounts, reductions or increases in the 'standard' amount. Finally there is a section on challenging decisions of a billing authority in relation to council tax.

WHO IS LIABLE

9.2 Liability to pay council tax accrues on a daily basis.[1] In deciding who is liable on a given day, 'it shall be assumed that any state of affairs subsisting at the end of the day had subsisted throughout the day'.[2] It is submitted that this provision does not require a focus on the situation at 11.59 p.m. The following remarks on similar provisions in LGFA 1988 are relevant here:

> where s67(5) refers to the state of affairs existing immediately before the day ends it is not requiring that attention be confined to the particular activities being carried on at a precise moment in time. What has to be considered is the use of the property with all its features, and all that s67(5) does is to identify the material time by reference to which any change in the use of the property is to be related.[3]

1 LGFA 1992, s2(1).
2 LGFA 1992, s2(2).
3 *Tully v Jorgensen (VO)* [2003] RA 233 at p241–242. See also *Mullaney v Watford BC* [1997] RA 225 at p232 which confirms that s2 is only relevant where it is alleged that there has been a change in the position as to liability.

9.3   It is possible for two or more persons to be jointly liable to council tax if they both satisfy the criteria for liability. This has an effect on the collection of unpaid tax,[4] but is otherwise straightforward. Special provision is made for married couples, people in civil partnerships and people who are cohabiting; where both are resident they are jointly and severally liable[5] even if on normal principles only one of them would be liable.[6] This rule does not apply where the non-liable partner is a student or suffering from a severe mental impairment, as defined.

The hierarchy of liability

9.4   The normal rules on liability are dealt with in this section. Specific provision has been made as to caravans and boats and as to the circumstances in which owners will be liable. These are discussed below.

9.5   LGFA 1992 imposes what is sometimes called a 'hierarchy of liability'. There are six categories of people who may be liable, designated (a) to (f). The person falling highest up the hierarchy will be liable. So if there is a person falling within category (a), that person will be liable, and anyone falling in categories (b) to (f) will not be. It follows that joint liability will arise only where two or more taxpayers both fall within the highest category to apply.[7]

9.6   The categories are as follows:

(a)   he is a resident of the dwelling and has a freehold interest in the whole or any part of it;

(b)   he is such a resident and has a leasehold interest in the whole or any part of the dwelling which is not inferior to another such interest held by another such resident;

(c)   he is both such a resident and a statutory, secure or introductory tenant of the whole or any part of the dwelling;

(d)   he is such a resident and has a contractual licence to occupy the whole or any part of the dwelling;

4 See the Council Tax (Administration and Enforcement) Regulations 1992, reg 27–28A.
5 That is, the full amount can be enforced against either of them. It is not a case of each being liable to pay 50 per cent of the total amount.
6 LGFA 1992, s9.
7 LGFA 1992, s6(3). If one or more of these taxpayers is a student or suffering from a severe mental impairment, he does not attract liability.

(e) he is such a resident; or

(f) he is the owner of the dwelling.[8]

9.7 In general, therefore, liability will fall on the resident who has the strongest title to occupy the dwelling. If there are no residents, it will in normal circumstances fall on the owner.[9] The two key concepts for understanding the hierarchy of liability are accordingly 'resident' and 'owner'. These are discussed below.

9.8 The definition of 'owner' can be dealt with shortly. 'Owner' is defined as follows in LGFA 1992:

> 'owner', in relation to any dwelling, means the person as regards whom the following conditions are fulfilled—
>
> (a) he has a material interest in the whole or any part of the dwelling; and
>
> (b) at least part of the dwelling or, as the case may be, of the part concerned is not subject to a material interest inferior to his interest;
>
> 'material interest' means a freehold interest or a leasehold interest which was granted for a term of six months or more.[10]

9.9 This definition, although somewhat convoluted, effectively speaks for itself. The owner is the person with the freehold or long leasehold interest, unless he has leased or sub-leased the property to another owner. A long leasehold interest in this context means a leasehold interest which was granted for a term of six months or more. Where a fixed-term lease expires, but rent continues to be paid, the leasehold interest created is a monthly periodic tenancy and therefore does not satisfy this definition.[11]

9.10 A person is a resident of a dwelling if he is 18 or older and 'has his sole or main residence in the dwelling'.[12] 'Residence' refers not just to a dwelling (as in the phrase 'a desirable residence') but to a dwelling in

---

8 LGFA 1992, s6(1)–(2).

9 This is subject to specific exceptions which are addressed under the heading 'Liability of Owners' at para 9.16 onwards.

10 LGFA 1992, s6(5)–(6).

11 *Macattram v LB Camden* [2012] EWHC 1033 (Admin).

12 LGFA 1992, s6(5).

which the taxpayer is resident; in other words, a dwelling in which he lives. It is therefore necessary to establish actual residence in a dwelling before considering whether that dwelling is the taxpayer's sole or main residence.[13] A dwelling will be the taxpayer's 'sole' residence if he does not reside anywhere else. The Court of Appeal has offered this guidance on identifying someone's main residence:

> 'sole or main residence' refers to premises in which the taxpayer actually resides. The qualification 'sole or main' addresses the fact that a person may reside in more than one place. We think that it is probably impossible to produce a definition of 'main residence' that will provide the appropriate test in all circumstances. Usually, however, a person's main residence will be the dwelling that a reasonable onlooker, with knowledge of the material facts, would regard as that person's home at the material time.[14]

9.11 Other formulations refer to the main residence as being the 'principal' or 'more important' residence,[15] 'the place where a person is habitually and normally resident apart from temporary or occasional absences of long or short duration', or the place 'where his home is, where he has his settled and usual abode, which he leaves only when the exigencies of his occupation compel him to'.[16] It follows that the time actually spent in each residence is relevant but 'only a factor, and certainly not a decisive one'.[17] Perhaps obviously, where someone sleeps is relevant to where he has his main residence.[18]

9.12 The application of these tests is a matter of fact and degree. Nevertheless, some assistance can be gained from looking at how this has happened in practice. The 'family home' has generally been considered to be the main residence above accommodation that is occupied for the purpose

---

13 *Williams v Horsham DC* [2004] EWCA Civ 39 at [23]–[26]. In both *R (Bennet) v Copeland BC* [2004] EWCA Civ 672 and *Parry v Derbyshire Dales DC* [2006] EWHC 988 the requirement for actual residence was found not to have been satisfied.

14 *Williams v Horsham DC* [2004] EWCA Civ 39 at [26].

15 *Frost (IT) v Feltham* [1981] 1 WLR 452, an income-tax case but accepted to be relevant to LGFA 1992.

16 *Bradford MCC v Anderton* [1991] RA 45.

17 *Bradford MCC v Anderton* [1991] RA 45.

18 *Mullaney v Watford BC* [1997] RA 225 at p231.

of work.[19] If there is a combination in respect of a dwelling of (i) wife and/or children resident, (ii) security of tenure (ownership or a long leasehold), and (iii) intention to return when possible, it may be that the only possible conclusion on the facts is that that dwelling is the main residence.[20]

9.13 Where a person's family lives is, for obvious reasons, treated as being of particular relevance when it comes to determining the location of his 'main residence'. A tribunal is entitled to infer, in the absence of any evidence to the contrary, that where a taxpayer has more than one residence, the main one is the one he shares with his wife and children.[21] Of course, the decision is a matter of fact in all the circumstances so the courts have acknowledged that a husband and wife may be happily married and yet have separate main residences, although noting that this is 'unusual'.[22]

9.14 A merchant ship plying the high seas cannot in law be someone's residence, and therefore cannot be his main residence.[23] A houseboat obviously can, as the specific provisions on liability for council tax in respect of moorings[24] demonstrate. The line between the two is likely to be determined by the fact that a mooring will not be a 'dwelling' at all unless there is a sufficient degree of permanence in the occupation of it.[25]

Caravans and boats

9.15 There are special provisions in LGFA 1992 for caravans and boats. These rules are simpler than the normal rules. If the owner is not resident, but someone else is, that other person will be liable. In any other

---

19 See *Ward v Kingston-upon-Hull CC* [1993] RA 71, *Codner v Wiltshire VCCT* [1994] RVR 169, *R (Navabi) v Chester-le-Street DC* [2001] EWHC (Admin) 796.

20 As happened in *Doncaster BC v Stark* [1998] RVR 80, a case about a serviceman who lived for most of the time on an RAF base. Nevertheless the High Court held that the only conclusion lawfully open to the tribunal was that his main residence was at the house he shared with his wife while on leave.

21 *Cox v London (SW) VCCT* [1994] RVR 171.

22 *Fowell v Radford* (1970) 21 P&CR 99, a case on similar provisions in the Leasehold Reform Act 1967. It may nowadays be rather less unusual for happily married couples to live separately, as more people follow the example once set by celebrities such as Helena Bonham-Carter and Tim Burton.

23 *Bradford MCC v Anderton* [1991] RA 45 at p58.

24 See para 9.15 onwards.

25 See paras 2.17 and 2.45–48 on the definition of hereditaments, and therefore of dwellings.

case, the owner will be liable.[26] The rules on joint liability are the
same as for normal dwellings. 'Resident' bears the same definition as
discussed above in relation to the normal rules for liability,[27] but the
definition of owner is modified somewhat to take account of the different legal arrangements (e.g. hire purchase) under which boats and
caravans may be owned.[28]

Liability of owners

9.16 LGFA 1992 provides a power to prescribe situations in which the
owner will automatically be liable, irrespective of the normal hierarchy of liability.[29] This power has been exercised in the Council Tax
(Liability for Owners) Regulations 1992. An 'owner' can be identified
according to the definition discussed above.[30]

9.17 These regulations provide for certain classes of dwellings in respect of
which the owner will be liable. The classes are as follows. For the most
part the regulations are self-explanatory as to which dwellings will fall
within which class.

9.18 Class A relates to registered care homes and hostels.

9.19 Class B relates to dwellings inhabited by a religious community, 'whose
principal occupation consists of prayer, contemplation, education, the
relief of suffering, or any combination of these'.

9.20 Class C relates to houses in multiple occupation, or 'HMOs'. It is the
one which has given rise to the largest number of court decisions.
Where it applies, the owner will only be liable if he has a freehold or
leasehold interesting in the whole of the dwelling.[31] It applies to the
following:

> a dwelling which
>
> (a) was originally constructed or subsequently adapted for occupa-
>   tion by persons who do not constitute a single household; or
> (b) is inhabited by a person who, or by two or more persons each
>   of whom either—

26 LGFA 1992, s7(1)–(3).
27 See para 9.10 onwards.
28 LGFA 1992, s7(7).
29 LGFA 1992, s8(1)–(3).
30 See paras 9.8–9.
31 Council Tax (Liability for Owners) Regulations 1992, reg 2A.

(i) is a tenant of, or has a licence to occupy, part only of the dwelling; or

(ii) has a licence to occupy, but is not liable (whether alone or jointly with other persons) to pay rent or a licence fee in respect of, the dwelling as a whole.

9.21 There are thus two separate conditions, either of which will qualify the dwelling as a Class C dwelling. They are commonly referred to as 'test (a)' and 'test (b)'.

9.22 Test (a) is satisfied if the dwelling was originally constructed or subsequently adapted for occupation by persons who do not constitute a single household. This requires a decision-maker to consider first whether the dwelling was originally constructed for that purpose, and if not, whether it has been subsequently adapted for that purpose.[32] The existence of security locks on bedroom doors has been held to be a sufficient adaptation to be capable of satisfying this test.[33] The intention of the person doing the work of construction or adaptation is not relevant.[34]

9.23 Test (b) relates to the actual occupants of the dwelling. It has two limbs. Limb (i) applies where each inhabitant is a tenant or licensee of only part of the dwelling. Limb (ii) deals with an alternative situation where each inhabitant has a licence to occupy the whole dwelling but is only liable to pay rent or a licence fee in respect of part thereof.[35] There is no need to show in either case that the occupiers 'do not constitute a single household'.[36]

9.24 The starting point in considering whether either limb of test (b) is satisfied should be tenancy or other agreements under which the dwelling is occupied. Unless there is evidence that those agreements are a sham, in the sense that they are dishonest documents which do not represent the true intentions of the parties, then they should be determinative.[37] That said, there may be 'exceptional circumstances' in which a document which appears to create a joint tenancy may be overridden.

---

32 *Hayes v Humberside VT* [1998] RA 37 per Kennedy LJ at p43–44.
33 In *Hayes v Humberside VT* [1998] RA 37. Insofar as *Pearson v Haringey LBC* [1998] RVR 252 appears to suggest otherwise, it can no longer be regarded as good law.
34 *Hayes v Humberside VT* [1998] RA 37 per Millett LJ at p44.
35 As summarised in *Naz v LB of Redbridge* [2013] EWHC 1268 (Admin) at [16].
36 *Hardy v Sefton MBC* [2006] EWHC 1928 (Admin) at [29].
37 *Watts v Preston City Council* [2009] EWHC 2179 (Admin) at [12] on (b)(ii), applied to (b)(i) by *Naz v LB of Redbridge* [2013] EWHC 1268 (Admin) at [45].

9.25   The main example of such circumstances is provided by *UHU Property Trust v Lincoln CC*.[38] This was a case in which the unusual facts lead the tribunal to conclude that, although the tenancy agreements under consideration were not a sham, they were nevertheless not determinative. It should be regarded as relating primarily to situations where 'property is let for occupation by those whose only source of income is housing benefit where often the landlord can have no expectation at the outset that they would ever be able to pay for more than the limited amount which the housing benefit gives, and for those cases in which the tenant is specifically allocated to a specific room', and/or to cases in which there is conflicting documentary evidence.[39]

9.26   Class D relates to dwellings occupied by domestic staff and their families which are, from time to time, occupied by the employer of the domestic staff. In essence this class prevents live-in domestic staff from being liable during periods when the owner is resident elsewhere.

9.27   Class E relates to dwellings which are inhabited by ministers of any religious denomination as a residence from which to perform their duties. It will be remembered that such dwellings are exempt while they are being kept empty awaiting the arrival of a new minister.[40] In the case of Church of England ministers in receipt of a stipend, the Diocesan Board of Finance rather than the owner is liable.[41]

9.28   Class F relates to dwellings provided to asylum seekers under the Immigration and Asylum Act 1999.

AMOUNT OF LIABILITY

9.29   LGFA 1992 contains provision regarding the setting of amounts of council tax for different bands of dwelling, and the addition of precept on behalf of other authorities.[42] These provisions are outside the scope of this book.

9.30   Once the amount per dwelling has been set, liability then arises on a daily basis in accordance with that amount.[43] This straightforward

---

38   [2000] RA 419. See also *Soor v LB of Redbridge* [2013] EWHC 1239 (Admin), which was also a case of conflicting documentary evidence: [32].

39   *Watts v Preston City Council* [2009] EWHC 2179 (Admin) at [13] and [21].

40   See para 8.53.

41   Council Tax (Liability for Owners) Regulations 1992, reg 3.

42   LGFA 1992, ss31–52ZY.

43   LGFA 1992, s10.

position as to liability is subject to the availability of discounts and also council tax reduction, which are discussed below. The basic liability may also in some situations be increased. Exemptions have already been discussed;[44] if a dwelling is exempt it does not generate a liability to council tax at all.

Reduced liability

9.31 In certain prescribed cases, ordinary liability to council tax is reduced. At present, two such reductions have been prescribed: in respect of disabled people and (in England only) in respect of annexes.

9.32 Annexes which would on normal principles form one dwelling with the main house will often fall to be 'disaggregated' under the relevant rules,[45] thus attracting an independent liability to council tax. In order to ameliorate the effect of these provisions, in England annexes in this situation benefit from a 50 per cent reduction in liability where one of two conditions are satisfied.[46] The first is that the annexe is being used by the inhabitant of the main dwelling as part of his or her sole or main residence. The second is that the annexe is the sole or main residence of a relative of the person who is liable to pay council tax in respect of the main dwelling.[47] 'Relative' is defined comprehensively in terms which include great-great-uncles and aunts.

9.33 A further or alternative reduction is available in either England or Wales where a dwelling is the sole or main residence of 'a person who is substantially and permanently disabled (whether by illness, injury, congenital deformity or otherwise)',[48] but only if it includes facilities provided for the needs of that person. Only the following facilities will qualify:

(1) a room used predominantly for meeting the disabled resident's needs (which is not a bathroom, kitchen or lavatory);
(2) an extra bathroom or kitchen;
(3) sufficient floor space for a wheelchair, where the disabled resident needs to use the wheelchair inside the dwelling.

44 See Chapter 8.
45 See para 8.11 onwards on disaggregation.
46 Council Tax (Reductions for Annexes) (England) Regulations 2013, reg 4.
47 Council Tax (Reductions for Annexes) (England) Regulations 2013, reg 3.
48 Council Tax (Reduction for Disabilities) Regulations 1992, reg 1.

9.34   In each case the facility must be 'essential or of major importance to his well-being by reason of the nature and extent of his disability'.[49] In order to fall within category (1), it is the use of the room itself that must meet this requirement; a living room will not qualify just because it is equipped with extra heating.[50] In other words, the room must be additional to the requirements of a person who is not disabled.[51]

9.35   To receive this reduction one of the liable persons must make an application to the billing authority each financial year.[52] Where the reduction applies, it has the effect of applying an alternative and lower banding to the dwelling than would normally apply; this will bring about a proportionate reduction in the level of liability.[53]

Discounts

9.36   The basic position on discount is as follows:

   (1) Dwellings with one resident attract a 25 per cent discount.[54]
   (2) Dwellings with no resident attract a 50 per cent discount.[55]

9.37   The amount of discount can be varied in certain cases, which are discussed below.

9.38   The definition of a 'resident' has been discussed above.[56] It excludes people under 18 and those who do not have their 'sole or main residence' at the dwelling.

---

49 Council Tax (Reduction for Disabilities) Regulations 1992, reg 3(2).
50 *Williams v Wirral BC* [1981] RA 189, a case on similarly worded previous provisions.
51 *South Gloucestershire Council v Titley* [2006] EWHC 3117, summarising the effect of *Williams v Wirral BC* [1981] RA 189 and also the intervening decisions to similar effect in *Sandwell MBC v Perks* [2003] EWHC 1749 (Admin) and *Luton BC v Ball* [2001] EWHC 328 (Admin).
52 Council Tax (Reduction for Disabilities) Regulations 1992, reg 3(1)(b).
53 Council Tax (Reduction for Disabilities) Regulations 1992, reg 4. Where the dwelling is already in Band A, there is a proportionate reduction in liability: reg 4(3A).
54 LGFA 1992, s11(1).
55 LGFA 1992, s11(2).
56 See para 9.10 onwards.

9.39 In calculating the number of people who are resident in a dwelling, certain classes of person fall to be disregarded. These are defined in Schedule 1 to LGFA 1992. In short summary, they comprise the following (the numbers in the list correspond to paragraphs in the schedule):

(1) certain detainees;[57]
(2) the severely mentally impaired;[58]
(3) those in respect of whom child benefit is payable;
(4) students satisfying prescribed conditions;[59]
(5) [this paragraph also relates to the definition of a student for these purposes];
(6) hospital patients;
(7) patients in care homes;[60]
(8) patients in private hospitals or care homes in Scotland;
(9) care workers meeting prescribed conditions;[61]
(10) residents of night shelters, homeless hostels and the like;
(11) other prescribed classes of persons, being:[62]

(a) members of certain international headquarters and defence organisations;
(b) members of religious communities;
(c) school and college leavers;
(d) persons with a relevant association with a visiting force;
(e) non-national spouses of students;
(f) certain diplomats.

57 See, in addition to the classes dealt with in LGFA 1992, sch 1, the Council Tax (Discount Disregards) Order 1992, art 2, which deals with persons in service custody pursuant to the Armed Forces Act 2006.
58 See the Council Tax (Discount Disregards) Order 1992, art 3 for a list of prescribed benefits.
59 See the Council Tax (Discount Disregards) Order 1992, art 4 and sch 1 and discussion at para 8.56.
60 And see further Council Tax (Discount Disregards) Order 1992, art 6 for the definition of a 'hostel' for these purposes.
61 The conditions are set out in the Council Tax (Additional Provisions for Discount Disregards) Regulations 1992, reg 2 and schedule.
62 These further classes are prescribed in the Council Tax (Additional Provisions for Discount Disregards) Regulations 1992, reg 3.

Modified discounts

9.40   The basic position on discounts may be modified in certain cases. Legislation allows billing authorities to make a 'determination' varying the amount of discount in certain prescribed cases; such determinations are to be made in respect of a financial year and can only be revoked or amended before the beginning of that financial year.[63]

9.41   The circumstances in which a billing authority may make a determination differ between England and Wales.

9.42   In England, there are three separate powers to make determinations altering the percentage discount which applies. Each applies to dwellings which fall into certain 'classes' as prescribed by Regulations.[64]

9.43   For dwellings in Class A or B, the billing authority may remove the discount altogether or reduce it, in all or part of its area.[65] These comprise all furnished dwellings which are not the sole or main residence of any individuals, whether or not occupation is restricted by a planning condition (i.e. second or holiday homes).[66] Caravan sites and moorings are not included in these classes.[67] Nor do they include dwellings where the person liable is also liable for another dwelling, for 'job-related' reasons. 'Job-related' reasons are narrowly defined, and are generally restricted to situations where there is a contractual or equivalent requirement to live in one of the dwellings.[68]

9.44   For dwellings in Class C or D, the billing authority may remove, reduce or increase the discount in relation to all dwellings in that class, or in relation to 'such description of dwellings of that class as it may specify in the determination'.[69] Class C relates to dwellings which are unoccupied (i.e. in which no one lives) and substantially

---

63   LGFA 1992, s11A(5), 11B(5), 12(5), 12A(7), 12B(8).
64   The Council Tax (Prescribed Classes of Dwellings) (England) Regulations 2003.
65   LGFA 1992, s11A(4) and the Council Tax (Prescribed Classes of Dwellings) (England) Regulations 2003, reg 3(1).
66   Council Tax (Prescribed Classes of Dwellings) (England) Regulations 2003, reg 4–5.
67   Council Tax (Prescribed Classes of Dwellings) (England) Regulations 2003, reg 6(1).
68   Council Tax (Prescribed Classes of Dwellings) (England) Regulations 2003, reg 6(2)–(3) and sch paras 1, 2 and 2A.
69   LGFA 1992, s11A(4A) and Council Tax (Prescribed Classes of Dwellings) (England) Regulations 2003, reg 3.

unfurnished.[70] Class D relates to vacant dwellings being structurally altered or which are the subject of major repair work to render it habitable, for a period of up to 12 months.[71] It will be remembered that in England, exemption has been removed from dwellings which are empty for less than six months and from dwellings undergoing works of refurbishment.[72] These classes effectively allow billing authorities to determine whether and to what extent they wish to grant discounts to make up for the loss of that exemption.

9.45   In Wales, the powers to alter the discount are rather simpler. A billing authority in Wales may remove or reduce the discount in relation to specified classes of dwellings throughout its area or in part of its area. The classes in question are Classes A, B and C, which correspond to those classes as defined in the English regulations discussed immediately above (although there are minor variations in the definition of 'job-related' dwellings).[73]

Council tax reduction schemes

9.46   Council tax benefit, which aimed to help low-income council taxpayers, has been abolished. In its place, each billing authority in England is required to make a 'council tax reduction scheme'.[74] In Wales, billing authorities are also required to make council tax reduction schemes, although regulations prescribe to a large extent what such schemes must include and not include.[75] The effect of these schemes is to reduce the council tax liability of those who fall within their terms.

9.47   Council tax reduction schemes in England must be targeted at those considered to be in financial need or persons in classes who are in general considered to be in financial need, even if not every member

---

70 Council Tax (Prescribed Classes of Dwellings) (England) Regulations 2003, reg 7 and 2(1).

71 Council Tax (Prescribed Classes of Dwellings) (England) Regulations 2003, reg 8.

72 See paras 8.41 and 8.47.

73 Council Tax (Prescribed Classes of Dwellings) (Wales) Regulations 1998.

74 LGFA 1992, s13A(2).

75 LGFA 1992, s13A(4) and the Council Tax Reduction Schemes and Prescribed Requirements (Wales) Regulations 2013, reg 11–15.

of the class is actually in financial need.[76] It follows that any criteria for reduction which is not related to financial need (such as a residence requirement) will be unlawful.[77] Criteria which relate to financial need but which discriminate unjustifiably against those with a protected characteristic (such as age, disability, race or religion) will also be unlawful.[78]

9.48 There is a statutory duty to consult major precepting authorities and also people who the billing authority 'considers are likely to have an interest in the operation of the scheme'. This applies when making[79] or revising[80] a scheme. Any such consultation must be fair. The courts have spelled out the following four requirements as a 'prescription for fairness':

> First, that consultation must be at a time when proposals are still at a formative stage. Second, that the proposer must give sufficient reasons for any proposal to permit of intelligent consideration and response. Third . . . that adequate time must be given for consideration and response and, finally, fourth, that the product of consultation must be conscientiously taken into account in finalising any statutory proposals.[81]

9.49 Furthermore, the courts have recognised that more specific information is required when consulting those who are not experts, and that fairness demands a higher standard of consultation where someone is

---

76 LGFA 1992, s13A(2).
77 *R (Winder) v Sandwell MBC* [2014] EWHC 2617 (Admin), where a scheme was declared unlawful on the basis of a two-year residence requirement.
78 *R (Logan) v Havering LBC* [2015] EWHC 3193 (Admin), in which the scheme was in fact held not to be discriminatory because it did include specific provisions relating to disabled people. The residence requirement in *R (Winder) v Sandwell MBC* [2014] EWHC 2617 (Admin) was considered by the court to be discriminatory, as well as being outside the scope of the powers granted by LGFA 1992.
79 LGFA 1992, sch 1A para 3(1); in Wales, the Council Tax Reduction Schemes and Prescribed Requirements (Wales) Regulations 2013 reg 17.
80 LGFA 1992, sch 1A para 5(5); in Wales, the Council Tax Reduction Schemes and Prescribed Requirements (Wales) Regulations 2013 reg 18.
81 *R v Brent London Borough Council, Ex p Gunning* (1985) 84 LGR 168 at p189, approved in this context by the Supreme Court in *R (Moseley) v Haringey LBC* [2014] UKSC 56 at [25].

being deprived of an existing benefit.[82] Taken together, these points mean that a consultation on one draft scheme must also invite views on the alternatives.[83]

9.50 A billing authority must have due regard for the need to advance equality of opportunity between those with a 'protected characteristic' and those without one.[84] This is known as the public-sector equality duty. The obligation is to have due regard in substance for the need to achieve the goals it sets out. It will be satisfied by a competent equality impact assessment, which need not be excessively detailed.[85]

Discretionary relief

9.51 In addition to the requirement to make council tax reduction schemes, and the various powers to determine what discount shall apply to various classes of dwelling, there is also a power to grant further or additional discretionary relief.[86] This may be granted in a specific case or by determining a class of cases to which it shall apply, and may reduce the liability to nil.[87]

Increased liability

9.52 The preceding parts of this section have been about powers to reduce the normal liability to council tax by a greater or lesser amount. In certain cases, however, billing authorities can in fact determine that council tax liability is *increased*. These provisions are intended to provide a mechanism to discourage certain forms of property ownership. The powers are more extensive in Wales than in England.

9.53 In England, a billing authority may determine that a 'long-term empty dwelling' shall not benefit from a discount and shall be subject an increase of up to 50 per cent in the normal liability that would apply.[88]

---

82 *R (Moseley) v Haringey LBC* [2014] UKSC 56 at [26].
83 *R (Moseley) v Haringey LBC* [2014] UKSC 56 at [29] and [39].
84 Equality Act 2010, s149(1).
85 The court rejected challenges to the application of the public-sector equality duty in this context in *R (Buckley) v Sheffield CC* [2013] EWHC 512 (Admin) and *R (on the application of Branwood) v Rochdale MBC* [2013] EWHC 1024 (Admin). See the latter at [63]–[65] for a helpful summary of the principles.
86 LGFA 1992, s1(1)(c).
87 LGFA 1992, s1(6)–(7).
88 LGFA 1992, s11B(1).

Such a determination overrides all other discount determinations that would otherwise apply.[89] A dwelling is a 'long-term empty dwelling' if it has been unoccupied and substantially unfurnished for a continuous period of two years, disregarding any breaks of up to six weeks.[90]

9.54 Certain classes of dwelling are protected from the possibility of such a determination being made. Dwellings in Classes E and F may not be made the subject of a determination increasing the liability to council tax.[91] Class E covers dwellings which are or would be the sole or main residence of a council taxpayer who is also liable to pay tax on a 'job-related' dwelling provided by the Secretary of State for Defence.[92] In essence this provision protects members of the armed forces who have no option but to reside elsewhere. Class F covers dwellings which are used together with another dwelling as part of a sole or main residence but which have been 'disaggregated' and so are treated as a separate dwelling.[93] This provision prevents dwellings from being subject to increased liability on the basis of what is effectively a technicality.

9.55 In Wales, the same power exists in relation to 'long-term empty dwellings' but with two modifications. First, the increase in liability can be up to 100 per cent.[94] Second, different rates of increase can be determined for different dwellings depending on how long they have been long-term empty.[95]

9.56 The classes of protected dwellings are also slightly different in Wales, comprising Classes 1 to 4.[96] These are as follows:

(1) dwellings which are being marketed for sale at a reasonable price or which have been sold but are awaiting completion. After a year, the dwelling ceases to fall within this class;

(2) dwellings which are being marketed for let on reasonable terms and conditions (including the rent), or where a let has

89 LGFA 1992, s11B(4). See para 9.40 on the making of 'determinations' generally.
90 LGFA 1992, s11B(8).
91 LGFA 1992, s11B(2) and the Council Tax (Prescribed Classes of Dwellings) (England) Regulations 2003.
92 Council Tax (Prescribed Classes of Dwellings) (England) Regulations 2003, reg 9.
93 Council Tax (Prescribed Classes of Dwellings) (England) Regulations 2003, reg 10. For the provisions on 'disaggregation' see para 8.11.
94 LGFA 1992, s12A(1).
95 LGFA 1992, s12A(2).
96 Council Tax (Exceptions to Higher Amounts) (Wales) Regulations 2015.

been agreed. After a year, the dwelling ceases to fall within this class;

(3) replicates Class F in the English provisions;[97]

(4) dwellings which would be the sole or main residence of someone if they were not residing in armed forces accommodation (whether as a member of the armed forces or a family member of such a person). This Class is therefore similar in effect to Class E in the English provisions, although its terms are subtly different and perhaps wider.

9.57 Both the powers so far discussed relate to empty properties, where there is a clear public interest in incentivising owners to bring them back into use. Second-home ownership is another controversial arrangement, which arguably leads to the underuse of property and causes affordability problems in rural areas. In England a billing authority's powers are limited to reducing the discount that such homes receive. In Wales, there is a further power to increase liability in respect of second homes.

9.58 This second power to increase council tax liability in Wales relates to 'dwellings occupied periodically'. A dwelling will fall within this description if there is no resident of it (i.e. it is not anyone's 'sole or main residence'[98]) and it is substantially furnished.[99] Such dwellings may be made subject to an increase in liability up to 100 per cent.[100]

9.59 Certain dwellings are once again outside the scope of this power. These include Classes 1 to 4 defined above. Classes 5 to 7 are also exempt from such a determination. They are as follows:

(5) dwellings comprising caravan pitches and boat moorings;

(6) dwellings which are subject to a planning condition preventing occupancy for a continuous period of at least 28 days in one year. Such conditions are routinely imposed on dwellings which are intended for use as holiday homes; as such there can be no objection to using them as second homes;

97 On which see para 9.54.
98 LGFA 1992, s6, and see discussion of this term at paras 9.8–9.
99 LGFA 1992, s12B(2).
100 LGA 1992, s12B(1).

(7) dwellings where the person liable for council tax is also resident elsewhere in a dwelling which is 'job-related'. 'Job-related dwellings' are strictly defined, in terms which generally require a contractual or equivalent requirement to live in the dwelling.[101]

DECISION AND CHALLENGE

9.60 The system of appeals is far more unified in respect of council tax than it is in respect of non-domestic rates. The Magistrates' Court has no role except in relation to enforcement.[102] Aside from certain matters which must be challenged by way of judicial review, all matters are dealt with by the Valuation Tribunal for England (VTE) or the Valuation Tribunal for Wales (VTW). This therefore includes:

(1) whether a dwelling is a chargeable dwelling or not;
(2) exemption from council tax;
(3) liability to council tax; and
(4) the amount of any liability (including discounts, reductions and increased liability).[103]

9.61 There is a two-stage procedure for challenging a decision of a billing authority on one of these matters. First, the person aggrieved by the decision must notify the billing authority of the fact that he is aggrieved by the decision, and the grounds on which he objects to the decision.[104] The billing authority then has two months to consider his complaint and respond. If it does not respond, or if the 'person aggrieved' is not satisfied by the response, he has a further two months

---

101 Council Tax (Exceptions to Higher Amounts) (Wales) Regulations 2015, sch 1.
102 On which, see Chapter 10.
103 LGFA 1992, s16(1).
104 LGFA 1992, s16(4)–(6). In the case of council tax reduction appeals in Wales, this must be done within one month of the decision. See the Valuation Tribunal for Wales Regulations 2010, reg 29 and Council Tax Reduction Schemes and Prescribed Requirements (Wales) Regulations 2013 sch 12 para 8(2) or the scheme prescribed in the Council Tax Reduction Schemes (Default Scheme) (Wales) Regulations 2013, sch 1 para 8(2).

to initiate the second stage and appeal to the VTE/VTW.[105] The time limits can only be extended if the VTE/VTW President is satisfied that the failure to meet the time limit arose by reason of circumstances beyond the appellant's control.[106] Regulations provide for the matters which must be included in the notice of appeal.[107]

9.62   On an appeal, the tribunal is able to remake the decision afresh. It is not confined to asking whether the billing authority's decision was reasonable or lawful. This even includes decisions about discretionary relief; the tribunal can exercise the discretion itself.[108]

9.63   The VTE has powers to correct accidental slips or omissions in their decisions.[109] A party to the decision who is unhappy with the decision can also request a review of the decision by the President.[110] He can review the decision if there has been some procedural irregularity in the proceedings, which includes a situation whereby the tribunal has clearly exceeded its statutory powers.[111]

9.64   The VTW has wider powers to review its own decisions either on procedural grounds or simply because 'the interests of justice' otherwise require a review.[112]

9.65   Otherwise, the tribunal's decision can be challenged by making an appeal to the High Court within four weeks.[113] This time limit can be

---

105 Valuation Tribunal for England (Council Tax and Rating Appeals) (Procedure) Regulations 2009, reg 21; Valuation Tribunal for Wales Regulations 2010, reg 29.

106 Valuation Tribunal for England (Council Tax and Rating Appeals) (Procedure) Regulations 2009, reg 21(6); Valuation Tribunal for Wales Regulations 2010, reg 29(5).

107 Valuation Tribunal for England (Council Tax and Rating Appeals) (Procedure) Regulations 2009, reg 20A; Valuation Tribunal for Wales Regulations 2010, reg 30.

108 *SC/CW v East Riding of Yorkshire Council* [2014] RA 279 at [19] to [25]; a decision by the President of the VTE so not technically binding.

109 Valuation Tribunal for England (Council Tax and Rating Appeals) (Procedure) Regulations 2009, reg 39.

110 Valuation Tribunal for England (Council Tax and Rating Appeals) (Procedure) Regulations 2009, reg 40.

111 *Reeves (VO) v VTE* [2015] EWHC 973 (Admin) at [65].

112 Valuation Tribunal for Wales Regulations 2010, reg 42. In the case of completion notices, the VTW can also review its decision on the basis of new evidence which was not reasonably available before.

113 Valuation Tribunal for England (Council Tax and Rating Appeals) (Procedure) Regulations 2009, reg 43; Valuation Tribunal for Wales Regulations 2010, reg 44.

extended by the High Court. Such an appeal is on a question of law only. The approach that the High Court will adopt has been explained as follows:[114]

(1) If the case contains anything which on its face is an error of law and which bears upon the determination, that is an error of law.

(2) A pure finding of fact may be set aside as an error of law if it is found without any evidence or upon a view of the facts which could not reasonably be entertained.

(3) An error of law may arise if the facts found are such that no person acting judicially and properly instructed as to the relevant law could have come to the determination under appeal.

(4) It is all too easy for a so-called question of law to become no more than a disguised attack on findings of fact which must be accepted by the courts. The nature of the factual enquiry which an appellate court can undertake is different from that undertaken by the Tribunal of fact. The question is: was there evidence before the Tribunal which was sufficient to support the finding which it made? In other words, was the finding one which the Tribunal was entitled to make?

(5) For a question of law to arise in those circumstances, the appellant must first identify the finding which is challenged; secondly, show that it is significant in relation to the conclusion; thirdly, identify the evidence, if any, which was relevant to that finding; and fourthly, show that finding, on the basis of that evidence, was one which the Tribunal was not entitled to make. What is not permitted is a roving selection of the evidence coupled with a general assertion that the tribunal's conclusion was against the weight of the evidence and was therefore wrong.

(6) An appeal court should be slow to interfere with a multi-factorial assessment based on a number of primary facts, or a value judgment. Where the application of a legal standard involves no question of principle, but is simply a matter of degree, an appellate court should be very cautious in differing from the judge's evaluation. Where a decision involves the application of a not

114 *Ramdhun v VTE* [2014] EWHC 946 (Admin) at [25]–[26].

altogether precise legal standard to a combination of features of varying importance, this will fall within the class of case in which an appellate court should not reverse a judge's decision unless he has erred in principle.

(7) Where the case is concerned with an appeal from a specialist Tribunal, particular deference is to be given to such tribunals, for Parliament has entrusted them, with all their specialist experience, to be the primary decision-maker. Those tribunals are alone the judges of the facts. Their decisions should be respected unless it is quite clear they have misdirected themselves in law. Appellate courts should not rush to find such misdirections simply because they might have reached a different conclusion on the facts or expressed themselves differently.

9.66 Decisions about individual dwellings or council taxpayers must therefore always be challenged before the VTE/VTW. LGFA 1992 provides, however, that certain matters which potentially affect a much wider array of dwellings and taxpayers can only be questioned by way of judicial review. This list of matters which may only be challenged by way of judicial review proceedings includes 'determinations' made by billing authorities in relation to discounts or increased liability and the making of council tax reduction schemes.[115] Perhaps by oversight, however, it does not include a determination that a certain class of cases should benefit from discretionary relief. Such decisions can therefore be challenged before the VTE/VTW.

9.67 Judicial review proceedings are taken at the High Court and must be commenced promptly and in any event within three months. A full discussion of the judicial review jurisdiction is outside the scope of this book, but the key point of distinction to note is that it does not allow the court to substitute its own judgment for that of the billing authority. The billing authority's decision will only be overturned if there has been an error of law. If it is overturned, it will then be for the billing authority to remake its decision in a lawful manner.

---

115 LGFA 1992, s66(2).

10

The collection of council tax

10.1 The framework for collection of council tax is set out in the CT Enforcement Regulations 1992.[1] In outline it is the same as that for non-domestic rates: service of a demand notice, or bill, followed if necessary by proceedings in the Magistrates' Court for a liability order, which itself may be enforced in a number of ways. Each of these three steps is dealt with in turn below.

10.2 A preliminary step, before council tax can be collected, is obviously to identify who is liable and for what amount. This must be done in accordance with the provisions discussed in the previous chapters. To assist it in this task, a billing authority has powers to request information from persons appearing to be residents, owners or managing agents of dwellings, and also from other public authorities.[2]

DEMAND NOTICES

Requirement for demand notices

10.3 The first step in collection is the service of a demand notice. It is an absolute requirement in respect of every council taxpayer in every financial year.[3] It is the demand notice which transforms the liability to council tax into a duty to pay.[4] A notice may relate to more than one year, but each notice may relate to only one dwelling.[5] This would

---

1 Council Tax (Administration and Enforcement) Regulations 1992.
2 Council Tax (Administration and Enforcement) Regulations 1992, reg 3, 4 and 12.
3 Council Tax (Administration and Enforcement) Regulations 1992, reg 18(1).
4 *Regentford Ltd v Thanet DC* [2004] EWHC 246 at [16].
5 Council Tax (Administration and Enforcement) Regulations 1992, reg 18(2)–(3).

suggest, by analogy with the position in respect of NDR, that a notice which relates to more than one dwelling cannot be relied upon to enforce payment of council tax.[6]

Timing of demand notices

10.4   Demand notices are to be served 'on or as soon as practicable after' the day the billing authority first sets a relevant amount of council tax. They can be served before the beginning of the financial year on the person who in the opinion of the billing authority will be liable.[7]

10.5   Assistance is to be gained in relation to the meaning of the word 'practicable' here from the cases dealing with non-domestic rating demand notices, where the requirement is very similar. The word 'practicable' in this context has been held to mean 'feasible' and 'possible to be accomplished with known means and known resources'.[8] Local authority resources will thus be relevant to what is 'practicable', but a 'billing authority will not be able to rely upon the suggestion that home-grown problems and inefficiencies rendered impracticable what would otherwise have been practicable'.[9] The issue in each case is by what point it was feasible for the billing authority to have discovered the identity of the taxpayer.[10] An alteration to the list will not excuse earlier failures in respect of the same hereditaments.[11] However, a billing authority's failure to notify the listing officer of information relevant to the alteration of the list will not invalidate a demand notice served after the list has eventually been altered, even if a ratepayer is prejudiced by that failure.[12]

---

6   *R (JJB Sports plc) v Telford and Wrekin BC* [2008] EWHC 2870 (Admin), which dealt with a prohibition in relation to non-domestic rates on demand notices including multiple years.

7   Council Tax (Administration and Enforcement) Regulations 1992, reg 17(2).

8   *North Somerset DC v Honda Motor Europe Ltd and others* [2010] EWHC 1505 at [63]–[64].

9   *North Somerset DC v Honda Motor Europe Ltd and others* [2010] EWHC 1505 at [64].

10   *Encon Insulation Ltd v Nottingham CC* [1999] RA 382.

11   *North Somerset DC v Honda Motor Europe Ltd and others* [2010] EWHC 1505 at [69].

12   *R (Secerno Ltd) v Oxford Magistrates' Court* [2011] EWHC 1009 (Admin).

10.6   This time limit is important, because if it is not complied with the demands may not be enforceable. A council taxpayer must show that he has suffered some procedural or substantive prejudice as a result of the late service of the notices.[13] If he can show this, then it will constitute a valid defence to recovery of the money whether in the Magistrates' Court or elsewhere.[14]

Contents of demand notices

10.7   The contents of a demand notice must accord with the provisions of the Council Tax (Demand Notices) (England) Regulations 2011 or in Wales the Council Tax (Demand Notices) (Wales) Regulations 1993. Failure to adhere to these regulations, however, need not result in the invalidity of the notice. Even if it does, there is specific provision in the regulations to the effect that where such failure was a mistake, amounts demanded in accordance with the CT Enforcement Regulations 1992 will continue to be payable.[15]

10.8   A demand notice before or during the relevant year (as almost all demand notices are) must require payments by instalments unless the ratepayer and the billing authority have agreed otherwise.[16] The number and nature of the instalments is set out in two 'schemes' in the schedule to the CT Enforcement Regulations 1992.

10.9   If instalments are not paid, a further notice is to be served demanding that they be paid.[17] If that notice is not complied with, or if further instalments are subsequently missed, the whole year's liability becomes

---

13 *Regentford v Thanet DC* [2004] EWHC 246 (Admin) at [21]. This means that the position in respect of both council tax and non-domestic rates is the same: see paras 5.4–5.

14 *Regentford v Thanet DC* [2004] EWHC 246 (Admin) at [22], *R (Hakeen) v VTS* [2010] EWHC 152 (Admin) at [26]; and see in respect of non-domestic rates *R (LB Waltham Forest) v Waltham Forest Magistrates' Court and Yem Yom Ventures Ltd* [2008] EWHC 3579 (Admin) at [46], *North Somerset DC v Honda Motor Europe Ltd and others* [2010] EWHC 1505 at [34].

15 Council Tax (Demand Notices) (England) Regulations 2011, reg 7; Council Tax (Demand Notices) (Wales) Regulations 1993, reg 5.

16 Council Tax (Administration and Enforcement) Regulations 1992, reg 21(1).

17 Council Tax (Administration and Enforcement) Regulations 1992, reg 23(1).

payable as a lump sum.[18] By contrast, a notice issued in respect of past periods must require payment as a lump sum.[19]

10.10 These provisions on what a demand notice must contain and the terms it must set for payment do not necessarily constrain a billing authority's freedom of action. A billing authority does not have to insist on payment according to the terms of the demand notice and can come to an arrangement with the ratepayer for payment in a different manner. Any such agreement rests purely on the billing authority's discretion not to take further steps of enforcement, however, and the ratepayer remains vulnerable to such further steps unless he pays in accordance with the terms of the demand notice.

10.11 It frequently happens that the liability is less than or more than was estimated when the demand notice was issued at or before the beginning of the relevant year. In any such case the billing authority is obliged to serve a further notice stating the revised amount and adjusting the amounts payable accordingly. If there is a further liability, this is due as a lump sum. If there has been an overpayment, this must either be repaid or credited to a subsequent liability of the council taxpayer.[20]

Service of demand notices

10.12 A failure to serve a demand notice correctly may mean that a court is prevented from ordering repayment of the sums demanded and/or may lead to a court order being set aside.[21] It is therefore very important for billing authorities to ensure that demand notices are served correctly. The provisions on service of council tax demand notices are practically identical to those in respect of non-domestic rating demand notices, discussed elsewhere.[22] If the name of the council taxpayer cannot be ascertained 'after reasonable inquiry', the notice can be served by addressing it to 'The Council Taxpayer' of the dwelling concerned.[23]

18 Council Tax (Administration and Enforcement) Regulations 1992, reg 23(3).
19 Council Tax (Administration and Enforcement) Regulations 1992, reg 20(4)–(5).
20 Council Tax (Administration and Enforcement) Regulations 1992, reg 24.
21 On which, see para 10.22 onwards.
22 See paras 5.12–17, The relevant provision for council tax is Council Tax (Administration and Enforcement) Regulations 1992, reg 2. There are no 'place of business' provisions in respect of council tax, for obvious reasons.
23 Council Tax (Administration and Enforcement) Regulations 1992, reg 2(3).

LIABILITY ORDERS

Final notices

10.13   Before commencing proceedings for a liability order in the Magis-
        trates' Court to recover sums due under a demand notice, the billing
        authority must serve a final notice allowing seven more days to pay
        the amount due.[24] There is no requirement to serve such a notice
        where notice has already been given of a failure to pay instalments,
        even if the proceedings are to recover the whole liability for that year
        and not just the missed instalments.[25] A final notice must be served
        in the same way as a demand notice.[26]

Complaint and summons

10.14   The actual proceedings in the Magistrates' Court are to be commen-
        ced by 'making complaint . . . requesting the issue of a summons'.[27]
        This is the standard method of instituting civil proceedings in the
        Magistrates' Court. In theory the complaint is made to the court
        by the billing authority and then a summons is issued by the court.
        In practice the two are usually generated as one document which is
        simply rubber-stamped by the court before being sent out by the bill-
        ing authority. The billing authority makes the complaint and so is
        known as the 'Complainant' in the court proceedings; the taxpayer is
        known as the 'Defendant'.

10.15   The summons must be served on the taxpayer. This is a vital step,
        because if the summons is not served correctly, it seems likely that the
        court will have no power to make a liability order.[28] A summons may
        be served on the taxpayer:

        (a) by delivering it to him; or
        (b) by leaving it at his usual or last known place of abode, or in
             the case of a company, at its registered office; or

---

24 Council Tax (Administration and Enforcement) Regulations 1992, reg 33(1).
25 Council Tax (Administration and Enforcement) Regulations 1992, reg 33(3).
26 On which, see para 10.12.
27 Council Tax (Administration and Enforcement) Regulations 1992, reg 34(2).
28 *Chowdhury v Westminster CC* [2013] EWHC 1921 (Admin) at [28], a case on
    similar provisions in relation to non-domestic rates.

(c) by sending it by post to him at his usual or last known place of abode, or in the case of a company, to its registered office; or

(d) by leaving it at, or by sending it by post to him at, an address given by the person as an address at which service of the summons will be accepted.[29]

10.16 'Last known place of abode', in this context, means the last place of abode known to the billing authority. This provision requires the billing authority to take reasonable steps to find out what the place of abode of the person to be served was, because 'knowledge' implies something more than a belief about where the last place of abode was.[30]

10.17 Where the summons is left at an address or sent by post, the deeming provisions discussed above will apply.[31] No liability order can be made unless 14 days have elapsed since the summons was served.[32] As such it would seem that the time of service is of importance and that, as a consequence, it is open to a council taxpayer to prove that a summons deemed served in the ordinary course of post has not in fact been received by him.

10.18 The limitation period for bringing liability order proceedings is '6 years beginning with the day on which [the sum] became due'. A sum does not become due until it has been the subject of a demand notice.[33]

Hearings in the Magistrates' Court

10.19 In considering whether to make a liability order, the Magistrates' Court 'shall make the order if it is satisfied that the sum has become payable by the defendant and has not been paid'.

10.20 This is exactly the same test as is set out in respect of non-domestic rates. The discussion on hearings in the Magistrates' Court in respect of those provisions is therefore relevant here. The only difference of substance relates to the fact that it is not open to a council taxpayer to

---

29 Council Tax (Administration and Enforcement) Regulations 1992, reg 35(2).
30 *R (Tull) v Camberwell Green Magistrates' Court, Lambeth Borough Council* [2004] EWHC 2780 (Admin) at [17]–[18].
31 See the discussion of the non-domestic rating provisions at paras 5.14, 5.22.
32 Council Tax (Administration and Enforcement) Regulations 1992, reg 35(2A).
33 See para 10.3.

dispute the existence or amount of his liability before the Magistrates' Court, as these are matters that can be raised on appeal to the VTE/VTW.[34] The court's role is thus more limited. Nevertheless, it has been suggested that 'the Magistrates' Court must enquire into questions as to whether the taxpayer is entitled to set off monies owed by the billing authority, or is entitled to say in law that the billing authority is precluded from asserting any liability to pay'.[35]

10.21 The position on costs awards[36] and appeals is the same as that in relation to non-domestic rates.[37]

Setting aside liability orders

10.22 The court retains a discretion to set aside liability orders if the following three conditions are satisfied:

(1) there is a genuine and arguable dispute as to the defendant's liability for the rates in question;

(2) the order was made as a result of a substantial procedural error, defect or mishap; and

(3) the application to the justices for the order to be set aside is made promptly after the defendant learns that it has been made or has notice that an order may have been made.[38]

10.23 These principles apply in the same way as they do in non-domestic rating cases.[39]

---

34 And the Magistrates' Court is therefore precluded from considering them: Council Tax (Administration and Enforcement) Regulations 1992, reg 57(1) and *Wiltshire Council v Piggin* [2014] EWHC 4386 (Admin) at [23], which resolves a previous conflict in the authorities on this.

35 *Hardy v Sefton MBC* [2006] EWHC 1928 (Admin) at [50], and see discussion at paras 5.31–33 which includes reference to earlier authorities to similar effect.

36 Governed in the case of council tax by the Council Tax (Administration and Enforcement) Regulations 1992, reg 34(7).

37 See para 5.34 onwards.

38 Set out in *R (Brighton and Hove City Council) v Brighton and Hove Justices* [2004] EWHC 1800 (Admin) at [31] and applied in numerous cases since.

39 See para 5.61 onwards for a full discussion. *R (on the application of Jones) v Justices of the Peace* [2008] EWHC 2740 (Admin) is an example of a case in which they were applied, after the introduction of the provision in the CT Enforcement Regulations 1992 dealing with the quashing of liability orders.

10.24   In addition, there is provision in the CT Enforcement Regulations 1992 for a liability order to be quashed on the application of the billing authority.[40] This power is wider than the common law discretion set out above in that it can be invoked whenever the billing authority considers 'that the order should not have been made'. If the court is satisfied that the order should not have been made, it must quash the order. It may also make a substitute liability order for a lesser sum if satisfied that such an order could properly have been made if the original application had been for the lesser sum. Such a substitute order must also include any costs from the initial order. There is no provision to include any further or different sum for costs in such an order.

Repayments

10.25   The CT Enforcement Regulations 1992 contain similar provisions on repayments to the relevant non-domestic rating regulations.[41]

FURTHER RECOVERY ACTION

10.26   In some cases, the making of a liability order may be sufficient to secure payment. In those cases where it is not, the billing authority has various options available to it. There is no limitation period in taking further enforcement proceedings once a liability order has been made.[42]

10.27   The familiar remedies of enforcement by taking control of goods,[43] commitment to prison[44] and insolvency are available. These are dealt with in the section on non-domestic rates.[45] The only relevant difference is in relation to commitment to prison – the court in a council

---

40 Council Tax (Administration and Enforcement) Regulations 1992, reg 36A.
41 Council Tax (Administration and Enforcement) Regulations 1992, reg 31 and see discussion of equivalent provisions in Chapter 7.
42 *Bolsover DC v Ashfield Nominees Ltd* [2010] EWCA Civ 1129.
43 Enabled by the Council Tax (Administration and Enforcement) Regulations 1992, reg 45 and LGFA 1992, s14(4).
44 Governed by the Council Tax (Administration and Enforcement) Regulations 1992, reg 47–48.
45 See Chapter 5.

tax case must be satisfied that the failure to pay 'which has led to the application' for commitment to prison is due to the taxpayer's 'wilful refusal or culpable neglect'.[46] This is different from the non-domestic rating provisions where the relevant failure is the failure to pay which led to the making of the liability order. The focus of a means inquiry in a council tax case will therefore be the period *after* the making of the liability order not the period *before* the making of that order as in a non-domestic rating case.

10.28   In addition, a billing authority has two further remedies available to it in respect of council tax: namely, charging orders and attachment of earnings orders. These are dealt with in turn below.

Charging orders

10.29   Where a liability order or orders have been made and £1,000 or more remains outstanding, the billing authority may apply to the local County Court for a charging order. The order will apply to the taxpayer's beneficial interest in the dwelling in respect of which he was liable to pay the tax. The costs reasonably incurred in obtaining the charging order can also be included in the order.[47]

10.30   The making of a charging order by the County Court is not automatic. The court must consider 'all the circumstances of the case' and, in particular, whether anyone other than the council taxpayer would be likely to be unduly prejudiced by the making of the order. The charging order may be made subject to conditions.[48]

10.31   A charging order can be varied or discharged on application by the billing authority, taxpayer, or any other person interested in the dwelling.[49] Whilst in force it can be protected by registration and has the same effect as an equitable charge created by the taxpayer.[50]

---

46   Council Tax (Administration and Enforcement) Regulations 1992, reg 47(2).

47   Council Tax (Administration and Enforcement) Regulations 1992, reg 50.

48   Council Tax (Administration and Enforcement) Regulations 1992, reg 51(1)–(2).

49   Council Tax (Administration and Enforcement) Regulations 1992, reg 51(4).

50   Council Tax (Administration and Enforcement) Regulations 1992, reg 51(3), (5).

Attachment of earnings orders

10.32 A billing authority may itself make an attachment of earnings order to recover amounts due under liability orders (plus some associated costs of enforcement).[51] The order can then be served on the taxpayer's employer, who is obliged to comply with it by deducting certain sums in accordance with the provisions in the CT Enforcement Regulations 1992.[52] No more than two such orders may be applied to a taxpayer at any time.

---

51 Council Tax (Administration and Enforcement) Regulations 1992, reg 37–44 and sch 3.
52 See Council Tax (Administration and Enforcement) Regulations 1992, reg 38.

Index

Note: Entries are listed by paragraph/note number.

Taylor & Francis eBooks

Helping you to choose the right eBooks for your Library

Add Routledge titles to your library's digital collection today. Taylor and Francis ebooks contains over 50,000 titles in the Humanities, Social Sciences, Behavioural Sciences, Built Environment and Law.

Choose from a range of subject packages or create your own!

Benefits for you

» Free MARC records
» COUNTER-compliant usage statistics
» Flexible purchase and pricing options
» All titles DRM-free.

REQUEST YOUR FREE INSTITUTIONAL TRIAL TODAY

Free Trials Available
We offer free trials to qualifying academic, corporate and government customers.

Benefits for your user

» Off-site, anytime access via Athens or referring URL
» Print or copy pages or chapters
» Full content search
» Bookmark, highlight and annotate text
» Access to thousands of pages of quality research at the click of a button.

eCollections – Choose from over 30 subject eCollections, including:

Archaeology	Language Learning
Architecture	Law
Asian Studies	Literature
Business & Management	Media & Communication
Classical Studies	Middle East Studies
Construction	Music
Creative & Media Arts	Philosophy
Criminology & Criminal Justice	Planning
Economics	Politics
Education	Psychology & Mental Health
Energy	Religion
Engineering	Security
English Language & Linguistics	Social Work
Environment & Sustainability	Sociology
Geography	Sport
Health Studies	Theatre & Performance
History	Tourism, Hospitality & Events

For more information, pricing enquiries or to order a free trial, please contact your local sales team:
www.tandfebooks.com/page/sales

 Routledge
Taylor & Francis Group

The home of
Routledge books

www.tandfebooks.com